Communicating Technical Information

Second Edition

Communicating Technical Information

A New Guide to Current Uses and Abuses in Scientific and Engineering Writing

Robert R. Rathbone

Professor Emeritus of Technical Communication
Massachusetts Institute of Technology

Addison-Wesley Publishing Company, Inc.

Reading, Massachusetts • Menlo Park, California • New York • Don Mills, Ontario
Wokingham, England • Amsterdam • Bonn • Sydney • Singapore • Tokyo • Madrid • San Juan

Library of Congress Cataloging in Publication Data

Rathbone, Robert R.
Communicating technical information.

Bibliography: p.
Includes index.
1. English language — Rhetoric. 2. English language — Technical English. 3. Technical writing. 4. Report writing. I. Title.
PE1475.R37 1985 808′.0666 84–20529
ISBN 0–201–06365–4

Cover design by Depoian/Taniuchi
Text design by Kenneth J. Wilson
Set in 10-point Palatino by Compset Inc.

ISBN 0–201–06365–4

FGHIJ-MU -89

Sixth printing, October 1989

Contents

Foreword

In the foreword to the first edition of *Communicating Technical Information,* I stated that the book had a modest function: to serve as a self-improvement guide for engineers, scientists, and technical writers and editors, whether on the job or in the classroom. The second edition has not abandoned that function but rather has widened it to meet more fully the important role of primary text for college and in-plant courses in technical communication. This new version thus retains the practicality and spirit of the old while filling a need for a textbook that concentrates on writing matters most bothersome to technical people.

To meet the new objective, the text is organized into three parts. Parts I and II present a logical sequence of material that lends itself nicely to classroom presentation. New chapters have been included on writing research proposals and achieving effective structure in reports. Part III contains two new chapters that cover material that many readers felt would enhance the usefulness of the book as a self-improvement guide: techniques and devices to aid the writer and word processing for the uninitiated.

Again, I have many people to thank for their advice and contributions — first and foremost my colleague and successor Dr. James Paradis, head of Technical Communications at M.I.T. I also wish to acknowledge the help of John Kirsch, John Kirkman, John Bennett, Jay Gould, and all those attending the M.I.T. summer programs on Communicating Technical Information who contributed their ideas on revising the book.

January 1985
Tuftonboro, New Hampshire

R.R.

Part I

General Problems

- *Getting started*
- *Organizing around a thesis*
- *Writing and revising*

Chapter 1

The Peaceful Coexistence

It isn't that engineers and scientists can't write — they just prefer to carry on a peaceful coexistence with the English language.

Observation by an M.I.T. student

Let's begin by being realistic: you wouldn't be reading this book unless you wanted some help with your writing. And although you may not be particularly enthusiastic about the project, the fact that you have made the step — even grudgingly — is important.

It is important because it is a start and because you have taken the initiative. It will not work wonders, however. You will have to do more than just read about writing — you will have to *write, write, write,* and you will have to rid yourself of any notions and habits that may have stymied you in the past.

This introductory chapter is intended to help you inspect yourself as a writer and to suggest ways to improve your approach before you begin your next writing assignment. The portraits of writer-types used for this purpose are caricatures; no writer could possibly possess all of the vices described!

You're really not that bad! If you have a lingering phobia about writing and hope that "something will turn up" so that you won't have to commit yourself, look at it this way. It's natural for you to dislike to do something if you feel you don't do it well. But there is no reason to believe that you aren't capable of becoming a competent writer. Engineers and scientists can produce effective writing if they have sufficient motivation and are given proper instruction. As proof, their journal articles on the whole are better written than those of many other professional people; their reports are even lucid compared with those of professional educators and psychologists.

What you must face, however, is that open hostility toward writing will harm you, and thus your career. You will have to learn at least to coexist with what you may now look upon as a necessary evil. Although computers are excellent tools for data and word processing, report writing still is an integral part of the human factor in every technical investigation. The researcher has to decide what needs to be said and to create the message. Peaceful coexistence, therefore, is better than limited warfare: you may soon find that relations improve as tensions disappear and you actually begin to like to write!

PLAN BEFORE YOU WRITE

To the outsider, it seems odd that engineers and scientists, schooled in an orderly approach to problem-solving, should have trouble with expository writing. There probably are numerous reasons for their difficulty, but I believe this to be basic: they seldom plan a writing assignment with the same care that they put into planning the technical end of a project. This is true even when the report that comes out of the project may be the only tangible evidence they have to show for their efforts.

If you identify yourself with this problem, try to divide your writing job into easy-to-handle tasks. It has been said, and rightly, that when you have something to write, the first thing to do, paradoxically, is to resist the urge to sit down and write! A writing assignment is a practical problem in communications. Its inputs are not bunched at one point in time, but occur throughout the life of the investigation. It therefore does not make sense to handle all of the writing tasks in one gigantic (and often exhausting) effort at the end. Many of these tasks can be accomplished more efficiently during the project, when information is fresh and motivation is high. And if you have access to a friendly computer, many also can be performed quickly and painlessly. Here's a random list:

- The audience can be identified and its needs established.
- The problem can be defined and the purpose of the investigation stated.
- The general organization and format of the report can be determined.
- Decisions can be reached on security classification and distribution.
- Information can be evaluated as it is gathered.
- Graphic aids can be roughed out.

- The bibliography can be prepared.
- An outline can be submitted to an editor or colleague for comment.
- Drafts can be written concerning work completed.

As you can see, most of these tasks represent decisions that you, the writer, must make before you begin to write. You will have enough to do later on.

AVOID BECOMING A STEREOTYPE

Technical writing is functional writing, not a form of fiction. Yet it often appears to the reader that some authors handle their papers as though they were writing a mystery story. They withhold a few pertinent facts, they include extraneous information, they report false leads, they build up undue suspense, and they leave questions unanswered. In short, some writers give the impression that readers should be made to work things out for themselves.

Another group of authors would lead us to believe that when a professional person writes something it must *sound professional* (whatever that is). Unfortunately, they interpret this to mean that technical writing cannot be natural, straightforward, personal, or easy to read. The pronoun "I" is outlawed, the passive voice supplants the active, the indefinite "it" occupies key positions in the syntax, and short sentences become long paragraphs.

Still others are firmly convinced that every writer is committed to producing a model of scholarly prose. A scholarly work deserves scholarly treatment. But the advocates of this line sometimes use style simply for style's sake. They choose words to impress rather than to inform. They include innumerable footnotes and references (some that are unnecessary, others that should be part of the text). They tailor the message to fit the syntax, not the other way around. They construct a massive bibliography. And before they know it, the whole thing becomes an exercise in one-upmanship rather than an act of communication.

And finally, there are those who believe that terseness is next to godliness. They have a special talent for brevity, but in their desire to be efficient they frequently sacrifice both clarity and readability. They mean well; they're earnest, hardworking people — liked by everyone except their readers. The following example depicts what happens to an art form (and could happen to your writing) when efficiency is carried too far. It is a satire of a report by a work study engineer, a specialist in method engineering, after attending a symphony concert at the Royal Festival Hall in London. The author, unfortunately, is unknown.

> HOW TO BE EFFICIENT, WITH FEWER VIOLINS
>
> For considerable periods, the four oboe players had nothing to do. The numbers should be reduced and the work spread more evenly over the whole of the concert, thus eliminating peaks of activity.
>
> All the twelve violins were playing identical notes; this seems unnecessary duplication. The staff of this section should be drastically cut. If a larger volume of sound is required, it could be obtained by electronic apparatus.
>
> Much effort was absorbed in the playing of demi-semi-quavers; this seems to be an unnecessary refinement. It is recommended that all notes should be rounded up to the nearest semiquaver. If this were done, it would be possible to use trainees and lower-grade operatives more extensively.
>
> There seems to be too much repetition of some musical passages. Scores should be drastically pruned. No useful purpose is served by repeating on the horns a passage which has already been handled by the strings. It is estimated that if all redundant passages were eliminated, the whole concert time of two hours could be reduced to twenty minutes and there would be no need for an intermission.
>
> The conductor agrees generally with these recommendations, but expressed the opinion that there might be some falling off in box-office receipts. In that unlikely event, it should be possible to close sections of the auditorium entirely, with a consequential saving of overhead expenses, lighting, attendants, etc. If the worst came to the worst, the whole thing could be abandoned and the public could go to the Albert Hall instead.

Economy of speech is to be commended. On the other hand, unless writers realize that *more words are necessary to convey a thought than simply to express it,* they are apt to indulge in false economy. Don't take "the obvious" for granted. Identify constants in formulas, specify the units in which data are presented, tell your readers what assumptions you are working under. Follow difficult passages with short summaries and provide transitional material that will take them smoothly from one idea to the next. You can read between the lines, but they can't.

OVERCOME YOUR INERTIA

For most people, getting started with any major piece of writing means overcoming human inertia, and unfortunately there is an abundant supply of it in all of us. One suggestion is to follow Newton's First Law: "A body in motion tends to remain in motion; a body at rest tends to remain at rest." Perhaps the best advice is "just start." Write something — anything that will put you in motion. You can throw the beginning away later if you wish.

Stephen Leacock, a writer long respected for his wit and wisdom by fellow writers and critics, put it this way:

> Suppose a would-be writer can't begin. I really believe there are many excellent writers who have never written because they could never begin. Many of them carry their unwritten books to the grave. They overestimate the magnitude of the task; they overestimate the greatness of the final result. A boy in a "prep" school will write "The History of Greece" and fetch it home finished after school. "He wrote a fine History of Greece the other day," says the proud father. Thirty years later the child, grown to be a professor, dreams of writing the history of Greece — the whole of it from the first Ionic invasion of the Aegean to the downfall of Alexandria. But he dreams. He never starts. He can't. It's too big.
>
> Anyone who has lived around a college knows the pathos of those unwritten books. Moreover, quite apart from the non-start due to the appalling magnitude of the subject, there is the non-start from the mere trivial difficulty of "how to begin" in the smaller sense, how to frame the opening sentences. In other words, how do you get started?
>
> The best practical advice that can be given on this subject is don't *start*. That is, don't start anywhere in particular. Begin at the end, begin in the middle, BUT BEGIN! If you like you can fool yourself by pretending that the start you make isn't really the beginning and that you are going to write it all over again. Pretend that what you write is just a note, a fragment, a nothing. Only get started.

Sometimes a change from using one method of transcribing your thoughts to another method will help to get you out of a rut. Perhaps you have heard some of your friends claim that using a cassette recorder solves the starting problem. I believe it does help many writers and recommend that you talk your way through an opening several times until you feel at ease with the technique. You may then want to go through a whole report. Be sure to have good notes to talk from so you won't have to search for things to say. Once you are no longer conscious of the recorder, dictation enables you to concentrate on the story. You have no time to worry about grammar and punctuation or to search for synonyms. Consequently, you transcribe your thoughts into prose quickly and with natural transitions. Later on, you can type this rough draft into a word processor and concentrate on polishing your organization and style. Chapter 13 gives you specific suggestions for doing this.

A curious but helpful suggestion for producing a rough draft of a long report — one that cannot be completed in a single sitting — is to stop in the middle of a section or a paragraph rather than at the end. When you resume the job (whether you are dictating or writing longhand), you then have enough starting energy built in to go right on without a struggle. Presumably, once you are in motion, new material will come faster and easier. Nonsense? Perhaps, but if your car battery were worn out, you wouldn't think it silly to park on a downgrade!

Of course, the biggest roadblock to getting started is not being ready to write — not knowing where you are going. If you know you have nothing to say, but insist on saying it, then you are beyond help. If, however, you believe you're ready but wish to check to be sure, ask yourself these questions:

- What was the objective of my investigation?
- Does the evidence I have gathered or the results I have achieved meet this objective?

If the answer to the second question is yes, jot down the main steps that connect the objective to the solution. You are then ready to fill in the details for an outline. If the answer is no, examine the reasons why you have not met the objective. Are they clear? Unbiased? Logical? Consistent? Credible? Reasonable? An unqualified yes here also means that you are ready to communicate your thoughts to others.

Finally, remember that not all investigations meet their initial objectives. A change in conditions, financial or otherwise, may dictate a change in objective. Perhaps this has happened to you and is the underlying cause of your difficulty in getting started. One way to handle the problem is to make a new outline of your report that focuses on the new objective. Pretend that the investigation started at that point. You can then describe the initial objective, and the reasons for abandoning it, as background material in your new introduction.

Chapter 2

The Wayward Thesis

If your writing falls apart, it probably has no primary idea to hold it together.

The Practical Stylist
Sheridan Baker

PROBLEMS, PROBLEMS!

Scientific and engineering investigations begin with a problem; most of the time they end with a solution. But the way is not always clear. Investigators frequently run into secondary problems that demand so much attention that they lose sight of the real reason for their involvement.

There is a story of an electrical engineer who was assigned the job of obtaining some data on pulse behavior in a computer circuit. Before he could begin his testing, he had to design special test equipment; this took a month. He then had to supervise the construction of the equipment and to check it out. These items took another month. To obtain the requested data, however, took only three days. In his report on the project, the engineer used thirty of the total of thirty-six pages to describe the design and construction of the test equipment. To him, this was the important information; he had spent practically all of his project time obtaining it. Yet this information was *not* of primary concern to the person receiving the report. As a result, the whole report had to be reorganized, material dropped, and the emphasis shifted to the solution of the original problem — the problem that brought about the investigation in the first place. The moral of this story is that much time may be wasted if writers do not first distinguish primary information from secondary information, and then shape their communication accordingly.

FIND A THESIS

Just as engineers or scientists need a hypothesis to guide them in conducting an investigation, so writers need a statement of thesis to guide them in *reporting* an investigation. They need something that will enable them to determine which course of action out of the many available to them is the best suited for a given communication.

The statement of thesis is a writer's tool. It is an internal communication that all writers should make to themselves before they begin the writing job. It is a mental communication, at least initially, and may be formed even before the investigation is completed. It tells a writer what the intent, general coverage, and emphasis of his or her communication should be. Without it, many writers suffer from "writer's block" or, as an editor friend of mine puts it, from "thesis paresis."

ANATOMY OF THE THESIS

A statement of thesis is a statement of "aboutness." It answers the question "What *about* the subject?" For instance, "Transistors replaced vacuum tubes because they require less power, occupy less space, and cost less" is one possible statement of the "aboutness" of transistors. "Although its cost is higher, process A produces a more reliable component than process B, and therefore should be adopted" concerns the "aboutness" of process A.

As the two examples show, a statement of thesis expresses an opinion. It is an appraisal of the significance of the subject. However, a writer does not necessarily have to express this opinion verbatim in the written text. Often it is preferable to let the facts speak for themselves.

Whether included as part of the text or not, the statement of thesis does represent what the writer hopes the reader will gather as being the central idea of the writing — what he or she hopes will be the main thought that readers will carry away with them when they finish the reading. Certainly there is a good chance of the reader getting the idea the writer has in mind if the writer forms the idea into a thesis first and then organizes the pertinent material around it.

The statement of thesis can also suggest how to organize the major elements of the subject matter in a report. Here is an example.

STATEMENT OF THESIS

The X-100 aircraft meets all of the USAF specifications for performance, structure, and instrumentation.
(Note: The statement answers the question "What about the X-100?")

GENERAL ORDER OF THE TEXT

I. *Performance:* Specifications, tests, results, evaluation
II. *Structure:* Specifications, tests, results, evaluation
III. *Instrumentation:* Specifications, tests, results, evaluation
IV. *Conclusion:* Synthesis of the evaluations

NEW REPORT: NEW THESIS

Many communications can spring from a single investigation, each with a different audience and a different intent. Thus coverage and treatment of the subject must be determined for each new reporting situation. For example, the first communication to come out of an investigation might be addressed to an audience interested primarily in test procedure and results; later a second communication might be issued to a different audience interested only in the economics of the project. In each case, a statement of thesis prepared in advance of the actual writing would tell the author what priority to assign to each segment of information he or she has collected. In effect, the statement of thesis serves the author as a filter for selecting which material is relevant for a given report and classifying it as primary or secondary. Illustration 2–1 depicts this important function.

Case Study

The following study illustrates one way of avoiding a wayward thesis.

Company X manufactures bulk chemicals. During the summer months, it has trouble protecting flammable liquids stored in its tank farm.

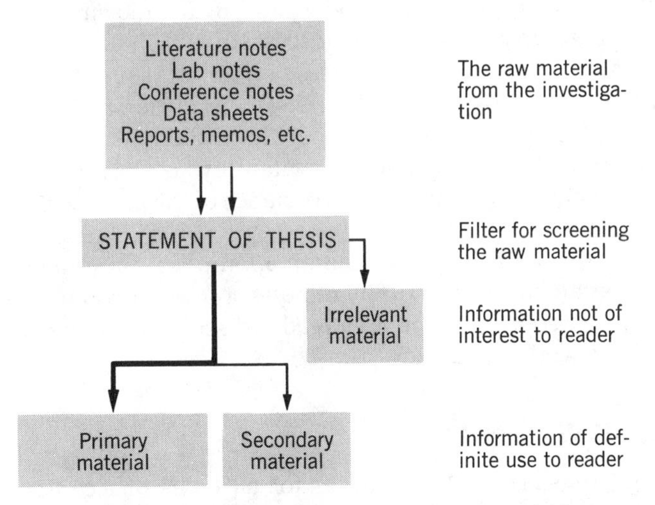

Illus. 2–1. Using the statement of thesis to screen material for a report.

All the tanks have standard protective devices, yet disastrous fires still occur.

Company Y manufactures an auxiliary system for protection of volatile liquids stored in above-the-surface tanks. You work for Company Y. You have talked to the management of Company X and are convinced that your equipment will solve their problem. Your assignment now is to write a report convincing them.

You begin organizing your thoughts by forming a tentative statement of thesis:

> Our equipment can solve their storage problem *because* it greatly reduces the fire hazards introduced by high atmospheric temperatures and electrical storms.

You examine this thought for a moment and realize that you will have to support it. So you revise the statement to include technical details:

> By introducing inert gas into the vapor space and by cooling the skin of the tank with water, our system will reduce by 90 percent the fire hazards caused by high atmospheric temperatures and electrical storms.

You decide that this version correctly represents the central idea you wish to present in detail as the main body of text. It expresses the aboutness of your protection system in relation to Company X's problem.

Your next step is to determine what information your readers will need if they are to accept your thesis. You realize that they will want to know immediately whether you really understand their problem. This thought prompts you to plan an introductory discussion of their equipment, the liquids they store, and the reported causes of their accidents. You also realize that they will wish to know what the costs of installation and operation will be; you decide, therefore, to follow your description of the system with a section on costs. These decisions on content make sense to you because your case depends on Company X's acceptance of the premise that the new system will work.

In summary, your thinking about an appropriate thesis and how you should support it has, in a very few minutes, provided you with a general outline of your report. You are now ready to screen your source material and to set up a working outline.

STATEMENT OF COVERAGE

It is not possible to state a thesis for all types of technical writing. Straight expository prose — writing in which no side is taken or no opinion offered — is better served by a *statement of coverage*. Operating instructions, descriptions of how to assemble something, and simple laboratory reports are typical cases.

The statement of coverage, like a statement of thesis, is a device to help the writer. It is formed before the writing begins so that it will provide a sense of direction. It is an informal statement, sometimes written out, sometimes not.

Unlike the statement of thesis, it is definitive rather than argumentative. It focuses on things themselves rather than on ideas about these things. It takes no line one way or another about the subject. It represents the author's opinion only as to how the subject is to be explained and how the explanation is to be organized.

A statement of coverage might read, for example:

> First, the theoretical operation of the system as a whole is presented in a logic diagram. The function of each segment is then discussed in detail to show how it interacts with its neighbors. This discussion leads to an analysis of the circuitry involved and the hardware necessary to implement it.

Except, perhaps, for the term, there is nothing new about a statement of coverage. But in their flush of enthusiasm to get the job done, inexperienced writers frequently bypass preparatory steps that experienced writers take automatically. Collecting one's thoughts about the coverage of the subject is perhaps the most important of these. Often this simple maneuver can save an otherwise worthless piece of writing.

Chapter 3

The Random Order

> ***Readers cannot understand information flung at them [randomly], a chip of geography here, a piece of economics there, a funny story tossed in for good (or bad) measure. Instead, they require that a subject first be reduced to some understandable scheme. . . .***
>
> *Informative Writing*
> John S. Naylor

Careful organization of material is the key to a successful scientific or engineering report. To the reader, clear and straightforward order reflects clear and straightforward thinking. A muddled report suggests a muddled investigation — whether it was so or not.

Organization is also important because it bears directly on timing. The order in which information is grouped determines when a reader will reach a specific chapter or section or passage or thought. For this reason alone, organization should never be left to chance.

For the purposes of this discussion, we shall examine organization at two levels: the general organization of the whole communication and the internal organization of the parts. The coverage will be limited to the formal report, since the requirements of organization, style, and format are more rigorous for it than for the informal type.

ANATOMY OF A FORMAL REPORT

Most formal technical reports have three distinct classes of material:

1. The front matter

2. The body
3. The end matter

The front matter consists of special service material. The following items are common:

The cover

The title page

The table of contents

The abstract

The foreword

Not all reports need a foreword, nor is a table of contents necessary if there are fewer than ten or twelve pages. A few reports substitute a letter of transmittal for a foreword. In such a case, the letter is included right after the title page.

The body is the business end of the report. It corresponds to the chapter coverage in a textbook and contains all the information necessary to satisfy the primary purpose of the communication.

The end matter usually contains a bibliography and an appendix (or appendices) of supplementary material of interest and help to some readers. Items that are commonly found in an appendix include the following:

- Tabulations of data (included in graphs in the body)
- Derivations of equations
- Sample calculations
- Sample forms used in the investigation
- Informative material for the secondary reader:
  - — Definitions of terms, descriptions of equipment
  - — Fold-out drawings

Appendix material should not be arranged in random order but in the order in which it is referred to in the text. Any item that the writer suspects will not be of use to the reader should be kept out, no matter how fond the writer is of it. Reference to appended material should be made at that point in the text where it will be of most use to the reader.

GENERAL ORGANIZATION OF THE BODY

The body of every formal technical report should have a clearly discernible beginning, middle, and end; this is the first rule of organization. This rule is not new; but many writers have found it extremely useful and it has survived.

One reason for its wide acceptance is that the three parts offer an easy framework within which to write. In a report on a research project, for example, the writer can give background information on the nature and objective of the project before he or she discusses the procedure followed, and can present the results before the conclusions and recommendations. This is the way most writers would tell the story if they were left to their own designs. The order is straightforward, and they feel comfortable with it.

The scheme also works nicely for reporting information that does not follow a general time base. In analytical reports, the beginning is the premise, the middle is the analysis, and the ending is the synthesis. In reports describing a new device or concept, the beginning is the overview, the middle is the presentation of details, the end is the summing up.

Because the three parts represent distinct communication functions, I prefer to call them by the following descriptive names: *the briefing, the evidence,* and *the evaluation*. The standard order in which they appear is represented in Illus. 3–1.

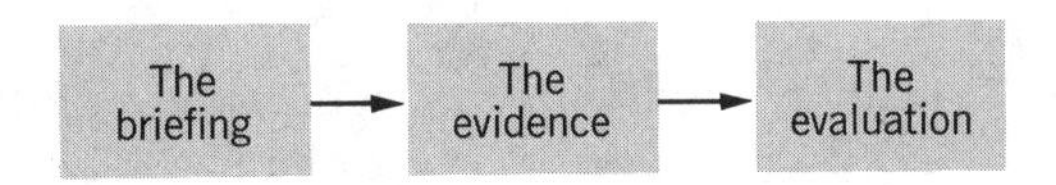

Illus. 3–1. Standard order of the parts of the body of a report.

The briefing constitutes the introduction. It supplies whatever preliminary information the reader needs in order to understand the evidence. It states the problem and the purpose and defines or explains any existing facts that have a direct bearing on the case. (Because introductions are so troublesome, they are treated in detail in Chapter 10.)

The evidence constitutes the objective reporting of the facts. It is the presentation of the evidence, pro and con, to the jury. Information is "out of order" if it does not bear directly on the premise, as announced in the briefing.

The *evaluation* constitutes an appraisal of the facts in terms of the purpose that was announced in the briefing. In the analogy of the court of law, it is the summing up by either the prosecution or the defense — depending on which side the writer represents. It may be followed by specific recommendations. (Not all reports have an evaluation section. If an opinion isn't asked for, a writer may decide not to offer any conclusions or recommendations, finishing the report with a short summary of the most important findings and leaving readers to draw their own conclusions. Or a report may simply supply the data for a particular use and no evaluation of the data may be necessary.)

Modification of the standard order is sometimes desirable. If the writer knows that the reader will be interested mainly in the evaluation, it could be moved into the middle position, forming an integrated unit with the briefing (see Illus. 3–2).

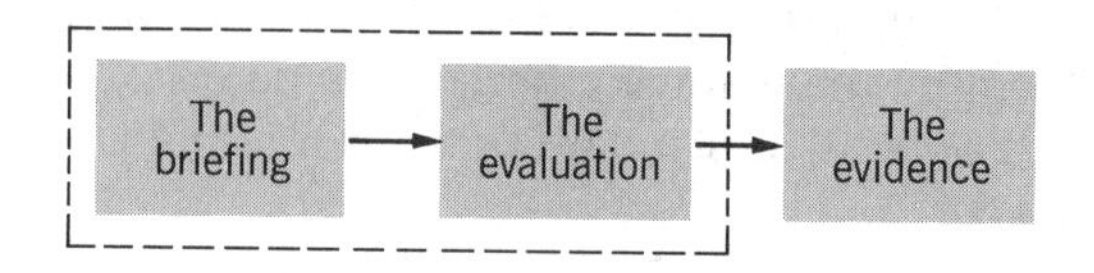

Illus. 3–2. Modification of the standard order.

Moving the evaluation into the first position, ahead of the briefing, is another possibility, but poses a danger. The writer has to be sure that sufficient introductory material is provided either in a foreword or letter of transmittal, or in the abstract. I have seen this modification used with moderate success in a few reports, but in each the need could have been met more efficiently if the writer had concentrated on providing a good informative abstract.

INTERNAL ORGANIZATION OF THE PARTS

Thomas Mann maintained that order and simplification are the first steps toward the mastery of a subject. He may not have had technical writing in mind when he said this, but I'm sure he would have agreed that the association is appropriate.

Once most readers begin a report, they look for signs to tell them how the parts are organized. If there is a table of contents, it will give them a quick map of where they are going. Then the in-text headings and subheadings take over. Such headings tell readers how far they have come along the route and what is to come; they show what material is primary, what is secondary, and how the parts are related; they provide transition; they help control the pace; they act as reference markers; and they add variety to the format. The following exhibit demonstrates how descriptive headings and subheadings help a writer achieve an orderly presentation.

Functional parts	*Headings used in report*
The Title	SELECTION OF CONTROL SYSTEMS FOR SPACE VEHICLES
The briefing (2 pages)	I. INTRODUCTION A. Problems of stabilization and attitude control B. Criteria for selecting control systems
The evidence (18 pages)	II. STABILIZATION SYSTEMS A. Gravitational-gradient stabilization B. Aerodynamic stabilization C. Radiation pressure stabilization D. Spin stabilization E. Comparison of stabilization systems III. ATTITUDE CONTROL SYSTEMS A. Control using gyros B. Control utilizing the earth's magnetic field C. Control using inertia wheels D. Control using reaction jets E. Control using reaction spheres F. Moving-mass attitude control G. Control by earth scanning H. Control by star or planet tracking I. Control using combined systems
The evaluation (2 pages)	IV. GENERAL EVALUATION V. RECOMMENDED AREAS FOR FURTHER STUDY

The exhibit also shows that (1) the presentation of evidence generally is much longer than the briefing or the evaluation, (2) each of the three functional parts, designated at the left, may contain more than one section, and (3) a carefully prepared outline could supply most of the headings and subheadings for a report.

Applying the three-part scheme to the organization of a report on a simple experimental investigation, we can assign the following general headings and subheadings to the body. (In an actual report, of course, the author would substitute words that describe the subject matter with which he or she is dealing.)

SPECIMEN HEADINGS FOR A REPORT ON AN EXPERIMENT

INTRODUCTION
Statement of problem
Statement of purpose

THEORETICAL ANALYSIS
Assumptions
Calculations

EXPERIMENTAL WORK
General procedure
Facilities and equipment
Tests
Results of tests

DISCUSSION OF RESULTS
Comparison of theoretical
  and experimental results
Conclusion

HOW TO HANDLE A TECHNICAL DESCRIPTION

Technical descriptions are difficult to write, but there are reliable rules to follow in planning the overall configuration. The most important of these is to PRESENT THE WHOLE BEFORE THE PARTS. In describing a new concept, a new device, or a new procedure, discuss the essence, the function, the purpose of the concept, device, or procedure — considered as a whole — before confronting the reader with the details.

Several years ago, I received a student report entitled "The Step-Down Detector." I was familiar with many types of detectors, but had never heard of this one and had no mental picture of what to expect. The report began with a list of equipment, followed by a circuit schematic. On page 3 I found this statement: "The step-down detector is a portable device to aid the blind. Basically, it consists of a light source, a photoelectric sensing element, and an alarm circuit that warns the carrier of the device that he is approaching a step down." If I had had this statement on page 1, I would not have wasted valuable reading time trying to guess the meaning of "step-down."

As you may have realized, not only does the whole-before-the-parts scheme meet an important psychological need of every reader but it also promotes sound structural organization. At the paragraph level, the whole becomes the topic sentence; at the section level, it becomes the introductory paragraph. And at the report level, it becomes the informative abstract.

Other important rules that will guide you in planning an effective technical description are:

1. Make sure that the manner in which the parts are organized is immediately apparent to the reader.
2. Try always to proceed from the familiar to the unfamiliar. Establish a common ground with the reader, either by direct reference or by analogy.
3. Determine whether your verbal description of "the whole" would be improved by the addition of a block diagram.
4. Check the pace. The most common mistake is to overcrowd sentences and paragraphs (see Chapter 7).

PUTTING THE PARTS IN ORDER

Look at a checkerboard in one light and you see black squares on red; in another light, red squares on black. From one angle, you may see columns running front to back; from another, rows running side to side or diagonally. Many patterns can emerge from a single graphic design.

So it is with a piece of writing. The writer might describe a device or a circuit in such a way that the reader sees a different pattern from the one intended. Or perhaps the picture is vague and a fuzzy pattern emerges.

No single way of assembling the bits and pieces of a technical description is best for all situations. The writer has to anticipate the angle from which the reader will view the writing and analyze the light in which it will be interpreted. Only through careful analysis of the subject and the reader can the writer determine what the best structure for the parts is.

The structural designs available to you as a technical writer are:

1. Expanded definition
2. Time relation
3. Space relation
4. Logical sequence
5. Enumeration
6. Cause and effect
7. Comparison and contrast
8. Classification or partition

Your first task is to select the design that will accommodate the subject matter and the audience you are to deal with. All designs should be given a chance to qualify. Sometimes a combination works best. Or sometimes the subject matter itself dictates your choice of design.

Once you have selected a design, you must then decide which pattern will be most likely to convey your thoughts successfully. Each design, like the checkerboard, offers alternative patterns; you should investigate all possibilities before deciding on one. Here are a few questions you should ask yourself before you begin to use any of the designs listed.

In an expanded definition, should you begin with the term or idea to be defined and then point out its distinguishing features, or should you begin with the features and lead up to the term or idea?

In a time relation, should you arrange your ideas from the past to the present, from the present to the past, or from present to past to present by a series of flashbacks?

In a space relation, should you go from right to left or from left to right? From top to bottom or from bottom to top? From inside to outside or from outside to inside?

In a logical sequence, should you proceed from input to output or from output back to input? Should you list the most important item first or last in a series based on order of importance?

In an enumeration, should you list items in the order of occurrence, order of importance, or order of familiarity?

In a causal relation, should you report the cause and then the effect, or the effect and then the cause?

In a comparison, should you develop a parallel comparison a step at a time, or should you present one of the items in its entirety before beginning the comparison?

In a classification scheme, should you group by structure, by principle of operation, by function, by cost, by weight, by size, by power, or by some other characteristic?

These questions show the many patterns available to you, the writer. But unless the design has been set for you, you must make the choice. If you are willing to experiment, to try alternative methods before settling on one, you will see definite improvement in your ability to communicate. The time spent experimenting will thus pay you dividends in clarity.

Postscript. Volta Torrey, past publisher of the M.I.T. alumni magazine, *Technology Review,* and a former editor of *Popular Science,* compares the reader's experience in beginning to read a technical report with the experience of someone entering a maze. Both are on their own in strange surroundings and both can easily be misled and become bewildered. The difference, of course, is that a maze is intended to confuse; it isn't considered much of a maze if it doesn't. A report, on the other hand, is intended to enlighten; it isn't considered much of a report if it doesn't.

So far as I know, no prize is being offered to the author of the best literary maze of the year. But there still are many competitors. Their point in common: an ability to mislead without even trying.

Chapter 4

The Faulty Structure

Write the way an architect builds,
who first drafts his plan and designs
every detail.

Schopenhauer

STRUCTURE AS AN ELEMENT OF STYLE

Broadly defined, "style" is the element in a piece of writing that distinguishes one writer's way of saying something from the way someone else might say the same thing. In so-called creative writing, this difference can be most pronounced. Many novelists, as you know, strive to develop personal styles and often are identifiable through their stylistic techniques, much as certain painters become associated with particular uses of color, tone, or brush strokes.

In technical writing, style also plays an important role, but in a different sense than in nontechnical prose. The writer's personality must not interfere with the clear and efficient transmission of a message. In this sense, then, a good style in technical writing is one that does its work quietly in the background without calling undue attention to itself. In no way is this an absence of style; rather it is one that centers on satisfying the reader, both psychologically and materially.

We usually think of choice of words as being the main ingredient of style. Words surely are important in this role, but so are the structures in which we place words: the paragraphs and the sentences. This chapter discusses how to use these elements successfully in technical writing. Word choice and usage are covered in the two following chapters.

PARAGRAPHS AS BUILDING BLOCKS

Although you probably have been told about the functions of paragraphs many times, I hope you won't mind another look. I will try to take you on a new tack by treating the paragraph as a unit of communication rather than as an element of grammar. If you accept the notion that paragraphs are the building blocks with which technical communications are built, the need to understand their functions and structure will then assume an appropriate importance.

Written communications are created when the entities we call topics are added together until the writer achieves the amount of detail necessary to communicate a given message to a given audience. In a well-structured piece of writing, each topic is represented by a paragraph and together the paragraphs form sections and the sections form chapters. Unfortunately, many writers do not take paragraph structure seriously. They write until they're tired and then break off for a new paragraph.

By itself, the well-structured paragraph is like a whole report in miniature, with an introduction (the topic sentence), a presentation of evidence (the developing sentences), and an evaluation (the concluding sentence). It therefore has unity — a primary specification, as you will see. It also has thrust, through the developing sentences, which advances the subject and keeps it on track. This specification is known as coherence. In a single-paragraph communication, the last sentence often presents a conclusion or serves as a summary. In a multi-paragraph report, the last sentence of a paragraph may also provide transition from its topic to that of the next paragraph.

Finally, a well-written paragraph indicates accurately what its relationship is to the subject of the whole paper through its topical relation to and its physical location under a subheading or a main heading.

Illustration 4–1 is a concise example of a paragraph serving as a whole report. When reading this passage, concentrate on the form and the technique.

> Modern weapon systems can be "exercised" without being fired. In the past, coastal defense guns were tested by actual firings at targets. But it is not advisable, nor practical, to test-fire Pershing II missiles once they have been deployed for actual defense. Instead, electronic test equipment injects simulated target information into a computer system which then checks the readiness of the weapons to respond. Actually, this simulated firing is, in one respect, a better test than the real thing: it exercises the system and trains the crew without expending the weapon.

Illus. 4–1. The paragraph as a whole report.

Note that this single paragraph has a conspicuous beginning, middle, and end. The first sentence announces the topic; the second, third, and fourth sentences develop the topic; and the last sentence evaluates the whole. Coherence is achieved in three ways:

1. By repetition of key words (or synonyms)
2. By use of words having the same base of reference
3. By use of words whose only function is to connect ideas

In the first category, the words "weapon," "exercised," and "fired" in the topic sentences are repeated in the developing sentences as "guns," "tested," "firings," "to test-fire," "missiles," and in the last sentence as "firing," "exercised," and "weapon." In the second category, the word "modern" is contrasted historically by the word "past" in the second sentence. In the third category, "but," "instead," and "actually" shift the emphasis in the development while keeping the thrust on track by connecting the thoughts of the individual sentences. I hope you also noticed that the last word in the paragraph ("weapon") relates back to the key word at the beginning, tying a bow around the package, so to speak. And finally, the paragraph illustrates that repetition can be effective without being boring or awkward, provided the writer is willing to use it carefully.

Illustration 4–2 is an excerpt from *How to Run Your Ford Car,* the owner's manual that came with every 1905 Model T Ford. (It's a prose classic!)

GO IT EASY

In the flush of enthusiasm, just after receiving your car, remember a new machine should have better care until she "finds herself" than she will need later, when the parts have become better adjusted to each other, limbered up and more thoroughly lubricated by long running.

You have more speed at your command than you can safely use on the average roads (or even on the best roads, save under exceptional conditions) and a great deal more than you ought to attempt to use until you have become thoroughly familiar with your machine, and the manipulation of brakes and levers has become practically automatic.

Your *Ford* car will climb any climbable grade. Do not, in your anxiety to prove it to everyone, climb everything in sight. A good rule is, if you crave the fame, climb the steepest grade in your neighborhood once, and let others take your word for it, or the word of those who witnessed the performance, for the deed thereafter.

Extraordinary conditions must be met when they present themselves — they should not be made a part of the everyday routine.

Illus. 4–2. Multiparagraph structure.

It not only illustrates good paragraph technique but also is easy and fun to read.

The first thing I want to point out in the example is that all four paragraphs relate directly to the heading GO IT EASY. Second, the idea of a "new machine" operated by a "new enthusiastic driver" is the thread that ties the paragraphs together. (Both have to "go it easy" until they are "broken in.") Finally, the single sentence at the end sums it all up neatly and forcefully. I hope you weren't bothered by the fact that three of the four paragraphs are single sentences. Each represents a separate topic, with the details developed by words rather than by sentences. Nothing would be gained by stretching them out. The author of this manual sensed that an informal, terse style was the best to use for instructing a proud and slightly awed owner, who probably still owned a horse and whose world had not yet become complicated with technical gadgets.

A final point related to all multiparagraph reports is that only paragraphs with one topic and with the topic sentence at, or near, the beginning give the reader the option of skimming. We all like to think that our readers should cherish every sentence we write. If we remember, however, how we act as readers, we will remember that the urge to skim is universal. Consider how deadly reading a report would be if it contained no paragraphing.

SENTENCE STRUCTURE

Individual sentences also contain their own logic and structure. Don't fight grammar; turn the tables and let it work for you. Analyze sentence structure for logic. Discard any rule that doesn't seem logical to you, but cultivate those that make sense.

You will improve your sentence structure if you recognize the elements that make up a sentence in their order of importance as communication devices.

1. *The Major Elements.* The major elements are *nouns* and *verbs* or their equivalents (pronouns and phrases, adjectives, and verb forms used as nouns). *A full idea can be expressed using only these elements.* They are the first words one learns in mastering a new language. Examples:

   John loves Mary.

   John loves her.

   Candy is dandy.

   To err is human.

Forgiving is divine.

Hitting the nail on the head isn't easy.

2. *The Supporting Elements.* These elements are the *adjectives* and *adverbs* or their equivalents (words, phrases, and clauses used as adjectives and adverbs). They qualify (modify) the major elements. *The major and supporting elements together can express ideas in detail.* Examples:

   Tall men admire short women.

   The new computer provides greater memory capacity than the older model.

   When you hit the nail on the head you're happy.

3. *The Service Elements.* Service elements work for the major and minor elements. They connect, separate, and compare and contrast these elements. By themselves, they do not express a meaningful thought. They are *conjunctions* and *prepositions* and can be single words or phrases or clauses. Examples:

   and, but, for, nevertheless, by, with

   on the other hand

4. *Minor Elements.* The minor elements consist of all the other types of words that we throw in for good measure. Many of them contribute little or nothing to the information we're trying to convey. Examples:

   Adjectives of degree — very, rather, somewhat

   The articles "the" and "a"

   Expletives — none are appropriate in technical writing

The first rule of thumb to follow is to *give priority in a sentence to major elements*. The beginning and the end of a sentence are the positions of most importance. These are the positions that attract the reader first. For example, if the subject of a sentence is at or near the beginning the reader is off to a good start. Use only those supporting elements that are necessary to convey *important* details. *Do not* bury the major elements under a pile of minor elements and supporting elements.

Remember, a short sentence is direct and easy to read. The major elements are seen immediately. It states thought without preamble or epilogue and it is preferred usage in technical communication.

A long sentence *per se* is not to be avoided. For one thing, it often avoids the choppy flow of a long series of short sentences. For another, it can provide a close connection between items that should be taken in thought together. Anything over twenty-four words should be monitored. Those that are developed on a time or procedure base usually are not difficult to follow provided the pace has been properly adjusted (see Chapter 7). Just keep in mind that a sentence generally becomes long when qualifiers are added. The question you must ask yourself is "Are there too many details for the reader to digest in the time it takes to read the sentence?"

Another important rule is *Put primary thoughts in primary grammatical structures; secondary thoughts in subordinate structures*. Primary structures are main clauses, subjects of verbs, objects of verbs, and the main verbs themselves. Subordinate structures are dependent clauses; objects of prepositions; and parenthetical words, phrases, clauses, and sentences. For example, in a complex sentence (one main clause and one or more dependent clauses) put the primary thought in the main clause, the secondary thought(s) in the dependent clause(s). Examples:

The results were achieved by field tests and are shown in Figure 4.
Comment: Subordination of the secondary thought (the figure reference) is not achieved with "and."

The results, shown in Figure 4, were achieved by field tests.
Comment: Subordination is achieved by using parenthetical structure.

A third sensible rule for sentence structure is *Shift from using the declarative sentence only if you gain emphasis, clarity, or ease of transition by doing so*. Don't be ashamed of the simplicity of the subject-verb-object sentence. It is useful because readers are familiar with this order, the thought flows naturally, and it has rhythm. Examples:

Declarative sentence: The quick brown fox jumps over the lazy dog.
Comment: Something does something. The emphasis is on the doer.

Inverted sentence: Over the lazy dog jumps the quick brown fox.
Comment: Emphasis shifted to the action. This sentence would be the right one to use if you wished to emphasize the jumping.

Periodic sentence: The loss of foreign markets, the drop in domestic sales, the deplorable state of our public relations, all stem from one source: inefficient management.
Comment: This type of sentence is an extension of the inverted sentence except that it usually is longer and is designed to build up suspense.

In the above example the reader cannot put together the sense of the sentence until the end. Try reading the sentence in standard fashion, beginning with "Inefficient management has caused . . ." You will soon see that the impact is not so great as in the original. (There, management really is clobbered.) The periodic sentence does not appear frequently in technical writing (and rightly so), but keep it in mind for occasions such as this: The turbid water, the silt, the offensive odor, and the absence of marine life are all caused by the run-off from the plant.

The last rule is *When the thought is complex, put it in a simple structure*. The logic here is: don't double the work for the reader. If a thought is complicated by many details or involved reasoning, do not put it into a long, complicated sentence. Break the thought into logical bits and put them into short sentences or into short clauses separated by semicolons. Most of the time two sentences will do the trick. If the bits happen to be parallel in thought, be sure to use parallel structure. The reader will have to decipher the structure only once. Examples:

Complex sentence: Thirty-six charpy blanks were cut from two bottom slices of each of six ingots, cooled to room temperature from 1759°F and reheated in salt at 1600°F to austenitize and refine the grain size.

Comment: The sentence crowds too many details into a single sentence. Without looking back, try to recall how many blanks were cut?

Revised sentence: Thirty-six charpy blanks were cut from two bottom slices of each of six ingots. All 216 were cooled to . . .

Comment: The new sentence allows the writer to clarify the total number of blanks by inserting a number after "all."

Complex sentence: Component A failed at 1000 hours of continuous operation in the system; the life of Component B, under similar conditions, was 1200 hours; however, that of Component C was 100 hours less than A's.

Comment: What a mess! Parallel thoughts are not in a parallel structure.

Revised sentence: Component A failed at 1000 hours of continuous operation in the system; Component B, at 1200 hours; and Component C, at 900 hours.

Comment: Straightforward organization and fewer words make this sentence much easier to read.

Chapter 5

The Artful Dodge

Any ambitious scientist must, in self-protection, prevent his colleagues from discovering that his ideas are simple . . . So if he can write his publications obscurely and uninterestingly enough no one will attempt to read them but all will instead genuflect in awe before such erudition.

Mathmanship
Nicholas Vanserg

In this age of scientific marvel it may be unfair (and even unpatriotic) to accuse scientists and engineers of double-talk or intentional hedging. But this is not to say that they never confuse or mislead when they write a report. The verbal haze commonly found in their writing can produce the same result unintentionally.

Unfortunately, confusion is not the only serious consequence; there is always the danger that the reader who is misled will think the writer is pulling an artful dodge just to save his own skin. Such a reaction usually spells disaster for the communication.

Fortunately, any writer can avoid giving the false impression of hedging by observing a few simple rules of common sense. This chapter discusses those rules.

THE VEILED INSULT

No writer wants to insult a reader, intentionally or unintentionally. Yet one of the quickest ways to achieve that result is to tell readers that something is *obvious* or *clear* — when to them it isn't obvious or clear at all. Repeated often enough, the ruse can succeed in frustrating readers completely.

The following examples represent typical cases:

> It should be obvious from the foregoing description that resonance is achieved through direct gamma-ray interactions.
>
> As we can clearly see, the factorial design makes very efficient use of the information in each run and reveals the presence of interaction between factors, each based on a variety of experimental conditions.
>
> The reaction of the input module is clearly demonstrated in the second set of equations.
>
> It is clear from the data that further investigation will be unnecessary.
>
> Although the tests were inconclusive, it is clear that the overall project was a success.

If something is obvious, there's no need to say so; *if there's a chance it isn't obvious*, why insult the reader by saying that it is?

Along the same line, some mathematicians like to save time by substituting the word "hence" for a page or two of intermediate calculations. This shortcut in itself isn't bad, but if the calculations are never included or are relegated to an obscure appendix without any mention in the text, readers quickly develop an inferiority complex when they try to bridge the gap. Don't force your readers to accuse you of deliberately trying to mislead them.

Sometimes writers are in such a hurry to finish a piece of writing that they can't resist shifting all responsibility for technical description onto graphic aids. "A picture is worth a thousand words," they remind themselves, and then quickly send the reader to Figure 1 for "details of how the system operates." Some graphic aids are intended to supplant words, but others can only supplement them. When both a verbal and a graphic description are needed, be sure to include both. Readers can quickly detect whether a drawing, photograph, or curve should be supported by text. Unfortunately, they can't do anything about it.

Finally, be careful about telling readers that something can be clearly seen in an illustration. Let them draw that conclusion. I remember writing that the controls of a certain piece of equipment could be *clearly* seen in a photograph. They were clear in the original glossy print all right, but not in the copy that finally appeared in the published report.

THE MEANINGLESS QUALIFIER

Qualifiers are designed to sharpen the words or expressions they qualify. When, however, the qualification cannot be visualized accurately or consistently, the audience can easily be misled.

Consider this statement: "The cost of the project should be *well under* a million dollars." How much is "well under"? $750,000? $10,000? $1.00? The qualification cannot be visualized in any meaningful way. (This is fine if you want to fool the audience, but not if you want to convince them.)

Suppose we reword the above statement: "The project will cost between $700,000 and $900,000." Although the dollar spread is wide, this revised statement is meaningful, and no reader would now accuse the writer of having an ulterior motive.

The expression "etc." can mystify readers for hours if it appears at the end of a series of unrelated items. Not knowing what comes next, they can only hope for the best. Suppose you received the following telegram:

> HOME BROKEN INTO LAST NIGHT — STOP — DOG, JEWELRY, MONEY, ETC. MISSING.

You probably would rush home in a hurry — worried, naturally, about your money, jewelry, and dog, but also concerned about the "etc." What could it be? Your doctoral dissertation? Your exercycle?

The mystery can be just as confusing in technical writing. "Etc." is meaningless if the reader cannot complete the missing items of a series:

> To build the component, we shall need a rag, a bone, a hank of hair, etc.

Be specific. Don't use "etc." to cover up laziness or lack of information or to mislead the reader into thinking that there are more items than there actually are. Reserve it for those instances in which spelling out the series would involve the reader unnecessarily:

> Match the pairs as follows: No. 1 with No. 7, No. 2 with No. 8, No. 3 with No. 9, etc.

The *floating comparative* is also a villain, yet may look innocent enough at a casual glance. It is the qualifier without a referent. We say that something is "faster" or "longer-lasting" or "more economical," but forget to complete the comparison. Thus we can have a legitimate referent in mind, but readers may think they are being given a sales pitch. Tie down the comparative and you will avoid this embarrassment.

The adverb "relatively" is another troublemaker. It is commonly used in such expressions as:

> a relatively high temperature or relatively expensive equipment.

The word is meaningless if the reader does not have a standard on which to base the qualification. It therefore must be defined for anyone not versed in the conventions of the subject being reported. "A relatively high temperature," for example, means one thing to someone working in pyrogenics; another thing to someone in cryogenics.

Other types of vague or meaningless qualifiers found in technical writing are:

a great many	considerably	very
more or less	mainly	rather
to some degree	reasonably	little
by and large	nearly	largely
a number of	appreciably	generally

After questioning one of my students as to what he meant by the term "reasonably sure," I received this amusing explanation: "'Sure' conveys no information by itself. It serves merely as a convenient point from which to express varying degrees of certainty. Each expression has special significance to the engineer." He then wrote the following list on the board.

Not quite sure

Fairly sure

Rather sure

Reasonably sure

SURE

Very sure

Real sure

Sure indeed

Sure as hell

In his book *The Elements of Style*, E. B. White calls the qualifiers *very, rather,* and *little* "leeches in the pond of prose, sucking the blood of words." I included them in the above list because these qualifiers are used to qualify other qualifiers, and thus can doubly confuse the issue. If we could see ourselves in print as we see others, we would never be caught saying:

a *very thorough* investigation	a *rather deadly* poison
a *very timely* occurrence	a *little more* difficult
a *rather dangerous* experiment	a *little less* frequent

If we jolted ourselves sufficiently, we might even see the wisdom of using an unadorned, unqualified phrase.

An IBM booklet on writing a technical paper offers the following passage to illustrate the evils of overqualifying. The unnecessary words are italicized.

> This study, *which is as yet inconclusive*, supports the theory, *within prescribed limits*, that the apparent, *though yet unmeasurable*, loss of iron from ferrite, *though in no other observable composition*, is due, *to the best of our knowledge*, to the atomization, *or what might thus be termed*, of Fe_2O on the cation points.

THE VAGUE REFERENCE

Most of you correctly follow one of the several accepted procedures for writing footnotes and compiling a bibliography, so these procedures will not be reviewed here. But you frequently violate the principles of documentation in statements that you make as part of the text. Here is an example of what I mean:

> *It is believed that* engineering colleges should hire more scientists as teachers.

This statement is both vague and dishonest. *Who* believes it? The impersonal "it" construction takes in the whole technical and academic community! Here is another:

> *The literature shows that* rain in Spain falls mainly in the plain.

Undoubtedly *certain* sources show this, but your readers are not helped if you casually cite all the literature as your authority. To them, you are taking unfair refuge behind a professional cliché.

And finally, here is a list of approaches, any one of which might easily destroy reader confidence:

> *The vast majority* of scientists feel that . . .
> *Comment:* A vague, wordy cliché
>
> These figures, *all from reliable sources*, indicate . . .
> *Comment:* A political runaround
>
> *A number of people* have asked why . . .
> *Comment:* Any number is "a number"; also, what people?

THE HIDDEN ANTECEDENT

In an article he wrote while head of the Mt. Wilson observatory, Paul W. Merrill suggested that an easy way to catch a reader off guard is to use a pronoun to refer to a noun that was mentioned a long way back, or to use a pronoun to stand for something not directly expressed. "If you wish to play a little game with the reader," Merrill continues, "offer him the wrong antecedent as bait; you may be astonished how easy it is to catch the poor fish."*

The pronouns that cause the most trouble in technical writing are "it" and "this" (and their plurals). They can easily become vague when the text contains two or more possible antecedents and the sense of the message does not tell the reader which one is the right one.

VAGUE "IT"

> If the gate brackets the signal of interest and if the amplitude of this signal has any frequency dependence, *it* will be recorded at the console.
>
> (Can you name the hidden antecedent?)

VAGUE "THIS"

> The preceding method provides an easy understanding of the so-called stress-relieving effect of closely spaced grooves and notches. *This* will now be illustrated.
>
> (What will be illustrated: the method or the effect?)

The authors of the two examples could easily have saved their readers from grief if they had repeated the appropriate nouns instead of using pronoun substitutes.

THE SEXIST "HE" OR "SHE"

Times and old customs have changed. So you would be wise to avoid using the pronoun "he" or "she" singly after "the reader," "the writer," "the editor," "the manager," and other nouns that specify a general title. There's no graceful substitute for the usage, but your best bet is to use the plural forms: for example, "writers . . . they." If you must use the singular form, "the writer . . . he or she" is preferable to the ugly "he/she." Never use "(s)he." If the worst comes to the worst, you may have to reshape your passage to eliminate any pronoun reference. For example, in the sentence "The writer can control noise if he exerts a little ingenuity," the "he" can be avoided by changing "if he exerts" to the phrase "by exerting."

* Merrill, Paul W., "The Principles of Poor Writing," in *The Scientific Monthly*, January 1947.

EUPHEMISMS AND "WEASEL" WORDS

Euphemisms are expressions substituted for ones that suggest something unpleasant, i.e., *pass away* for *die*. They are popular with writers when failures or setbacks have to be reported or when the results of an investigation are not as expected. Students writing lab reports, for instance, are all too familiar with the old saw:

The experiment reflected uncertain results.

(*Meaning:* The experiment failed.)

The practice is not restricted to the classroom, however. The following description from the *Style Manual* of the Aerojet-General Corporation* points out that professional people often succumb as well.

REPORTING FAILURES OR SETBACKS

One of the most difficult tasks for the technical writer is the adequate, yet prudent, discussion of disappointing results obtained in a program.

The most common mistake is to attempt to gloss over failures or setbacks. It is natural for an engineer to be optimistic about his work and to minimize difficulties. However, this tendency must not prevail to the extent that the reader is misled (or thinks that an attempt is being made to mislead him). Nothing can destroy a reader's confidence more effectively than the feeling, justified or not, that the writer has tried to deceive him. Any attempt to belittle difficulties will create a bad impression. The report writer will do better to present a frank discussion of the difficulty and then proceed to a logical analysis, followed by a discussion of the steps taken to deal with the problem.

The following examples of "weasel" words would not be likely to impress a critical reader favorably:

1. Three tests were conducted and all firings were very successful except the third, during which a malfunction occurred . . . damaging the unit beyond repair.
2. Six tests were run and the firing curves were very smooth for all except the first, third, fourth, and sixth.
3. . . . a pressure surge occurred. A search for the missing parts was initiated.

* Reprinted by permission of Aerojet-General Corporation, El Monte, California 91734.

4. . . . a major nonfiring occurred.
5. . . . divergent instability developed.
6. . . . a rapid structural failure occurred.

It is also possible to carry frankness too far, as in the following example:

> Although instruments of different manufacture, but comparable quality, from those discussed herein might be substituted, to do so would entail, in some instances, extensive changes in this report.

HEDGE WORDS

Close cousins to euphemisms and weasel words are hedge words — words and phrases that writers feel let them off the hook, so to speak. Hedges come in all shapes and sizes and are excellent for any writer who wishes to develop a noncommittal style.

The amateur begins with the common "possible," "probable," or "perhaps," but soon graduates to the more sophisticated "it would seem that," "it is likely that," or "the data indicate that."

The professional knows no bounds:

> *So far as we know,* the design specifications for the X-100 are more rigid than those for the X-99.
>
> The results are as reliable *as the time and equipment allotted to the project would allow.*
>
> All the evidence *seems to point to the conclusion that* a renewal of the contract will be necessary.
>
> The data *may be considered reliable until such time as new evidence is uncovered.*

Also, if you think it advisable to include some such phrase as "more research is needed before we can provide . . ." in your summary or conclusion, take care! This type of qualifier can easily have a negative connotation. When uttered repeatedly, it often becomes an annoying hedge in the minds of readers, or at best, projects a tone of indecisiveness. Readers have a low tolerance of anything they believe to be an evasion of responsibility.

But, you say, you like your job and want to protect it. What then? Well, you can avoid going out on a limb and still satisfy the reader if you will observe this simple *and reasonable* rule:

At the outset, state clearly all the specifications, bounds, limits, and assumptions you had to follow in conducting your project and by which you wish it to be judged. Have done with qualifying at once, then state each result, conclusion, and recommendation in as straightforward a manner as you can. You will not be accused of executing an artful dodge.

Chapter 6

The Ubiquitous Noise

The difference between the right word and the almost-right word is the difference between "lightning" and "lightning bug."
Mark Twain

In information theory, the term "noise" is broadly defined as any factor within or outside of a communication system that alters the intended message; that is, the thoughts the originator wishes to convey. I am by no means an expert in this field; nor do I know how the human brain functions. My purpose in these paragraphs is simply to show that written communication is a very complex operation, even when attempted by the best of writers.

All human communication systems are susceptible to noise, some systems more than others. But none seems to suffer so much from the effect as writing.

The odds are high against writing a noise-free message, because the English language is at times so irrational and because, in the traditional mode of communicating that most of us still use, the originator and the receiver have no feedback channel of any kind between them. The reader has no way to influence the transmission and the writer has no way of knowing that noise exists.

In a simple communication situation consisting of a writer, a report, and a reader, the elements of the system are as shown in Illustration 6–1. The writer performs four tasks: he or she gathers information, analyzes it until it is understood, encodes it into language symbols, and transmits the symbols as written signals. The reader reverses the process: he or she reads the signals, decodes them into symbols, interprets the symbols into information, and analyzes the information until it is understood.

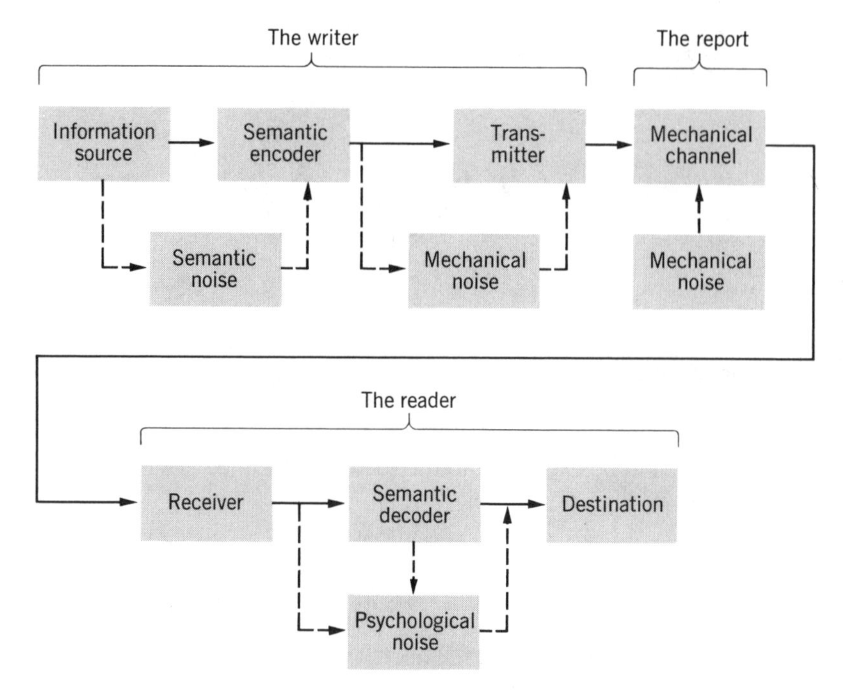

Illus. 6–1. Elements of a simple communications system.

Noise can, and does, occur at each of the three elements: the writer may unwittingly create semantic noise at the encoder or mechanical noise at the transmitter; the report itself may be subject to mechanical noise from numerous other sources; and the receiver is often the victim of psychological noise, the source of which may be the message itself, semantic or mechanical noise in the message, or some external stimulus.

SEMANTIC NOISE

To encode a message in writing, the originator must translate thoughts into word signals and assemble the signals into message units, such as phrases, clauses, sentences, and paragraphs. To communicate successfully, he or she must select signals that the reader will be able to retranslate into the original thoughts and must use a structure that will show the precise relationship the thoughts have to one another. Failure in either of these operations will create "semantic noise."

Every writer does have some control over semantic noise. But most readers do not expect the amateur writer to be able to control it completely at all times. They *do* expect a favorable signal-to-noise ratio, however — one that permits them to read through the interference easily and to restore the original message. The following techniques, conscientiously applied, are certain to reduce semantic noise to a tolerable level.

Select the right word. The right word is the word that will convey a given thought accurately, clearly, and efficiently to a given audience. What is right for one set of readers, however, may not be right for another. Whenever you have a choice among words, you must base your definition of "right" on the reader's background and knowledge.

Define the specialized word. If a notion has but one word to represent it, the problem is not in word choice but in whether the specialized word needs to be defined. Some technical words are in such common use that they do not need to be explained to the reader. "Voltage," for example, is a technical word, but the nontechnical person understands it. "Acetylcarboxypeptidase," on the other hand, is known only to a biochemist. Such a word represents just a large hole in a sentence until the reader knows what it means.

Make the context clear. Many technical words (as well as nontechnical words) have several meanings. The verbal context in which these words are used must tell the reader which meaning is intended. "Program," for example, might confuse the reader if it were used first for its special meaning in computer work and then in the next sentence for its common meaning as a plan of procedure. I can recall how much astonishment the term "bus driver" (describing a computer component) caused when it appeared in a paper addressed to a nontechnical audience!

A small point, really not worth arguing about, but isn't it time we stopped using the term "practicing engineer"? The usage may be consistent with the notion of "practicing" a licensed profession, but the term as we generally apply it is not restricted to licensed or registered engineers. Also, the double meaning of "practicing," to me at least, is awkward. Why not follow the sciences in this regard? There are no "practicing scientists," since the sciences are not licensed professions, but there are as many kinds of scientists as there are kinds of engineers, and the scientists get along very nicely without "practicing"!

Prefer the plain word. Plain words create less noise than ornate, formal words because they are known to more people. If a formal word is the exact symbol of what you wish to convey and its informal counterpart is not, then use the formal word — defining it when necessary. But if the plain word represents the image equally well, or better, always prefer it.

Why say this . . .	*if you mean this?*
utilize	use
terminate	end
magnitude	size
optimum	best
unique	uncommon
conjecture	guess
necessitate	need
fabricate	build

H. W. Fowler, in his *Dictionary of Modern English Usage*, points out that many of our words are "not the plain English for what is meant . . . but translations of these into language that is held more suitable for public exhibition." He calls these words "indoor words" and "outdoor words," the indoor words being "those that the mind uses in its private debates to convey to itself what it is talking about." He insists, however, that most of the time the outdoor words are not needed, and suggests that the less change from the indoor to the outdoor word the better. His reasoning makes sense: the writer thinks of "try" but translates it into "endeavor"; the reader sees "endeavor" but retranslates it into "try."

Be wary of fad words. Most fad words are short-lived. Their potency decays rapidly and those that manage to survive become clichés. A case in point is the fad popular ten years ago for adding the suffixes "-wise" and "-ize" to form new words (the meanings of which more often than not were nebulous). Some still are used by weather forecasters and sports announcers on radio and TV, however, to the dismay of most listeners:

"Weatherwise, the outlook for Boston is rain."

"To finalize the forecast for today, here again is Fred Windy."

"Teamwise, the Rams have sustained too many injuries."

And although "-wise" and "-ize" have lost favor, a new coinage, using "-ware" as the suffix, has appeared. It was created by a computer company that expanded its production to include books, which they proudly advertise as "bookware." Heaven help us! "Textware" and "novelware" will be next.

Prefer the single verb to the verb-noun combination. Long verb phrases can create noise by distracting readers from what is significant and by slowing them down. A weak verb-noun combination is a poor substitute for a single strong verb.

Why say this . . .	*if you mean this?*
to make a study of	to study
to arrive at an approximation as to how much	to estimate
to take into consideration	to consider
to give some assistance to	to help
to have a particular preference for	to prefer
to conduct an investigation of	to investigate

In their book, *Federal Prose, or How to Write in and/or for Washington*, Masterson and Phillips give some beautiful examples of how a wordy verbal construction can lead to pompous diction:

English	*Federal prose*
Scissors cut.	Scissors effect scission, functionally.
Hens lay eggs.	Gallinaceous ovulation is effected only by hens.
Haste makes waste.	Precipitousness entails negation of economy.

Place modifiers near the words they modify. If you don't, the reader might think they modify something else and miss your point entirely.

In the neighborhood of the crystal's resonant frequency, Pierre Vigoreaux has derived the following values for the parameters. (Those Frenchmen do get around!)

The student who cheats in the final analysis harms nobody but himself. (What about the one who cheats in the preliminary analysis?)

The baseball commissioner opened a hearing on charges that the fiery Baltimore manager struck and kicked an umpire at Memorial Stadium half an hour ahead of schedule today. (Possible, knowing the temperament of some managers.)

The effect of thickness of surface film on friction, metal transfer, and wear has been described as being related to the plastic roughening mechanism in a paper by one of the authors. (Sandpaper?)

When properly plugged into a circuit, you would expect this meter to indicate the amount of current. (If you could still read, that is.)

The actual weight of the impellers will vary from that calculated by 50 to 150 percent. (Calculated by 50 percent?)

The object of this thesis is to design a solar still that will convert sea water to drinking water on a portable scale. (In bottles, cans, or on tap?)

Having swollen because of the rain, the workman was unable to remove the bracing. (Call an ambulance!)

Eliminate roundabout expressions. Words that say things indirectly are always potential sources of noise. A complete list would take up pages. Here are a few of the more common ones.

Wordy	*Concise*
due to the fact that	because
on account of	because
in the event that	if
in case	if
a large number of	many
a great deal of	much
at the present time	now
despite the fact that	although
for the purpose of	for, to

I have already condemned the vague "it is believed that . . ." in Chapter 5 and would like to indict the rest of the "it" clan now. You will always improve a sentence by dropping the "it" construction.

Original: *It is essential that* the design specifications be modified.
Revision: The design specifications *must* be modified.

Original: *It is recommended that* the design specifications be modified.
Revision: The design specifications *should* be modified.

Original: *It is apparent that* the design specifications were modified.
Revision: *Apparently* the design specifications were modified.

or

The design specifications were modified.

Original: *It can be shown by tests that* the higher the speed the greater the rate of error.
Revision: *Tests show that* the higher the speed the greater the rate of error.

Original: *It can be seen in Figure 4 that* the pulse repetition frequency does not remain constant.
Revision: The pulse repetition frequency does not remain constant (see Figure 4).

or

Figure 4 shows that the pulse repetition frequency does not remain constant.

Taboo: *"It is obvious that . . ."* and synonymous phrases.
"It is felt that . . ." and synonymous phrases.
"It is needless to say that . . ." (Why say it, then?)

Subordinate secondary ideas. You have been reminded in Chapter 4 to put primary thoughts in primary grammatical constructions and secondary thoughts in secondary constructions. Violating this rule of logic is one of the main contributors to noise. We err mainly by placing secondary thoughts in primary constructions, but could readily avoid this if we make a practice of checking the importance of the thought every time we use an "and" to connect it to its neighbor.

"And" is a coordinating conjunction; that is, it connects items of equal weight. When used to connect clauses, it announces to the reader that both the clause preceding and the clause following are independent clauses — primary grammatical constructions. If both the thoughts contained in these clauses are not primary, the "and" has been used improperly. The following sentences illustrate this point.

Secondary thought not subordinated: The results have been summarized in Table 1 and prove that the earlier method was less effective.

Secondary thought subordinated: The results, summarized in Table 1, prove that the earlier method was less effective.

In the second version, the reference to the table has been relegated to a secondary construction, shifting the emphasis of the sentence onto what the results prove.

MECHANICAL NOISE

Mechanical, or physical, noise assaults the reader from two sources — from the report itself and from the environment in which the report is read. Most forms of environmental noise are well known: loud conversation in an adjoining room, telephones ringing, interruptions by visitors, construction equipment operating outside. Environmental noise, of course, adversely affects all kinds of human activities. But it is especially troublesome in written communication because the person responsible for the success of the communication, the writer, can do nothing to prevent its occurrence.

Mechanical noise within the report itself is another matter. The writer is accountable for it, and can control the worst of it by exerting a little ingenuity and effort. Some of the common forms are given below, with comments on how to lessen their intensity.

Poor-quality copy. This form of noise is serious, but fortunately it occurs infrequently. The main source: stretching the capability of a particular duplicating system in order to meet a deadline. Writers can help by getting their copy to the typist (or publications group) on time. Poor copy also sometimes results from duplicating on both sides of a page. Watch out for "burn-through."

Overcrowded text. The eye needs white space on a page to relieve the monotonous tramp of solid black lines. Large blocks of text, if continued page after page, produce an adverse psychological effect as well as an adverse physical effect. Variation in paragraph length, indentation, use of headings and subheadings, change in spacing between lines when the structure changes, generous margins on both sides of the text, and vertical listing of points all help to break the monotony and to provide a resting place for the eye.

Inadequate left margin. Many reports are held together by a stiff binding of some sort. If the left margin of text is less than one inch, some of the message can become concealed in the binding. Insist on a one-and-a-quarter-inch left margin, even if your report is to go into a loose-leaf notebook. (Otherwise a three-hole punch may cut into the text.)

Binding too stiff. It is always desirable (and sometimes necessary) for readers to be able to keep a report open to a certain page without holding it with their hands, turning it upside down on their desks, or applying weights. Losing one's place by having a book or report close suddenly is not a quieting experience. Shop around for covers and bindings; try different types before you order in quantity.

Inadequate labeling of figures and tables. Figures and tables should bear labels that identify their parts, without need of reference to the text. Readers frequently thumb through a report, stopping at the graphic illustrations; they are enlightened only if the illustrations are self-sufficient.

You can diminish noise from this source if you: (1) assign a number to each graphic aid; (2) provide a main title; (3) label all axes, columns, and rows; (4) specify the units in which data are presented; (5) define in a note all special symbols and abbreviations; (6) make curve lines bolder than grid lines; and (7) use as open a grid as will still permit the reader to interpret the data correctly.

Late reference to figures and tables. Attention should be directed to a graphic aid the moment its information is required. Late reference forces readers to reread the text in order to appraise it in terms of the new evidence. If a reference comes very late, readers may give up before ever reaching it.

Regarding placement, figures and tables should be located after the reference in text, not before. Nothing is more disconcerting to readers than to come upon an illustration and have no idea how it fits into what they are reading or to feel they have missed the reference to it.

Too few headings. Headings act as signposts: they help the reader skim, locate material, and follow the organization of the message while reading along. Without headings, the reader has to rely entirely on paragraphs to accomplish these tasks.

Most inexperienced writers err on the side of too few headings rather than too many. Anything can be overdone, but, in general, headings are effective antinoise devices.

Typos, spelling mistakes, punctuation errors. These lapses in the mechanics of a piece of writing create noise, sometimes serious noise. I recently found the following typographical error in a technical report: "value" for "valve." It was caused by poor handwriting on the draft (the typist mistook the word) and it changed the entire sense of the passage.

The most common spelling mistakes involve homonyms (sound-alikes). The worst offenders are:

Principle (noun "rule," "basic law") for *principal* (adj. "chief," "foremost")

Forward (verb "to transmit") for *foreword* (noun "front word")

Effect (verb "to cause to happen") for *affect* (verb "to influence")

Continuous (adj. "unbroken") for *continual* (adj. "closely repeated")

Discreet (adj. "prudent," "modest") for *discrete* (adj. "consisting of distinct elements")

Punctuation errors are serious if they create ambiguity; bothersome if they mislead even momentarily. Consistency in usage precludes many from developing. The surviving errors have to be caught during proofreading. In particular, watch for hyphen omissions in compound adjectives, as in these examples:

Ambiguous: "A remote control station" (Remote control or remote station?)

Specific: "A remote-control station" (A station for "remote" control; the station might be nearby)

Ambiguous: "A short delay component" (Short component?)

Specific: "A short-delay component" (Component that produces a short delay)

A plan of attack. You can wait for proofreading to catch the routine mechanical errors in a manuscript, such as typos, punctuation omissions, and faulty page numbering, but you should not wait to go after weaknesses in format or physical structure. Before you write the draft, determine how the readers will wish to use your finished product. If you are writing a field manual, will it help them if you specify a size that will fit in a pocket? Should the printing be made large enough to be easily read in poor light or while they are performing some physical task? If you are compiling a reference book, would the readers be helped if the main sections were tabbed and many subheadings used? In your instruction book, would a single fold-out drawing or separate in-text drawings be better for the intended use? Do you anticipate revisions? If so, would a loose-leaf binding be more practical than a stiff one?

The more mechanical matters you settle early in the writing project the more time you will have later on to concentrate on reducing semantic noise. Procrastination itself breeds noise.

We can also introduce noise when we use the language of mathematics. This is admirably illustrated in the following short satire from *Chemical Digest*, published by Foster D. Snell, Inc.

> Although most of our lives at Snell are spent making complex problems into simple answers for our clients, sometimes we amuse ourselves doing the opposite. We feel that our colleagues in engineering, physics and chemistry spend too much time complicating the simple before they publish their scientific papers. They do this in a great many ways, but the principles can be easiest illustrated by a simple mathematical example. Every paper to be published must above all conform to certain basic precepts whether it is clear or not. Every new chemist or engineer must learn early that it is never good taste to designate the sum of two quantities in the form:
>
> $$1 + 1 = 2 \qquad \text{(I)}$$
>
> Anyone who has made a study of advanced mathematics is, of course, aware that:
>
> $$1 = \ln e$$
>
> and that:
>
> $$1 = \sin^2 x + \cos^2 x$$
>
> further:
>
> $$2 = \sum_{n=0}^{\infty} \frac{1}{2^n}$$

Therefore, equation (I) can be expressed more scientifically in the form:

$$\text{In } e + (\sin^2 x + \cos^2 x) = \sum_{n=0}^{\infty} \frac{1}{2^n} \tag{II}$$

This may be further simplified by use of the relations:

$$1 = \cosh y \cdot \sqrt{1 - \tanh^2 y}, \qquad e = \lim_{z \to \infty} \left(1 + \frac{1}{z}\right)^z.$$

Equation (II) may therefore be rewritten:

$$\text{In} \left\{ \lim_{z \to \infty} \left(1 + \frac{1}{z}\right)^z \right\} + (\sin^2 x + \cos^2 x) = \sum_{n=0}^{\infty} \frac{\cosh y \cdot \sqrt{1 - \tanh^2 y}}{2^n} \tag{III}$$

or:

$$\text{In} \left\{ \lim_{z \to \infty} \left(1 + \frac{1}{z}\right)^z \right\} + (\sin^2 x + \cos^2 x) - \sum_{n=0}^{\infty} \frac{\cosh y \cdot \sqrt{1 - \tanh^2 y}}{2^n} = 0. \tag{IV}$$

At this point, it should be obvious to even the casual glance that equation (IV) is much clearer and more easily understood than equation (I). Of course, there are various other methods which could have been employed to clarify equation (I), but these should become obvious once the reader has grasped the underlying principle.

After carefully studying this brief discussion, the neophyte should be better able to appreciate the advantages of these methods when he chances upon them in his reading of the literature. When he has once acquired some facility in the application of these notations, he will be able to write technical reports and papers for publication, the clarity and simplicity of which will surely inspire comment.

But of course we never use these methods in writing reports to our clients.

— Philip A. Crispino*

* Reprinted by permission of Foster D. Snell, Inc., New York.

Chapter 7

The Neglected Pace

The most dramatic situation, the lightest humour, the profoundest wit, the most illuminating revelation of truth, will fail to reach its mark if the proper pace has been misjudged.

Reginald O. Kapp

Undeniably, pace is an important variable in conveying technical information. But it cannot be defined in quantitative terms, nor can it easily be separated from other communication variables and examined in the laboratory.

In oral communication, pace is the rate at which the speaker presents information to the listener. It is not so much the speed of speaking as it is the timing of the important ideas. The speaker paces the talk so that when these ideas are presented the audience is ready for them and will give their full attention.

In written communication, pace is the rate at which the *printed page* presents information to the reader. It therefore is not directly determined by the speed at which the writer writes (although careful writing often promotes more rapid reading), but it does influence the speed at which the reader reads. As in oral communication, pace controls timing.

The proper pace in technical writing is one that enables the reader to keep the mind working just a fraction of a second behind the eye as he or she reads along. It logically would be slow when the information is complex or difficult to understand; fast when the information is straightforward and familiar. If the mind lags behind the eye, the pace is too rapid; if the mind wanders ahead of the eye (or wants to), the pace is too slow.

Every reader likes to be able to finish a report as quickly as possible, but a rapid pace does not guarantee rapid comprehension. The writing must be paced so that the reader will be able to understand the content without having to stop or reread. When a slow pace achieves this timing better than a rapid pace, the reader achieves a better overall reading time.

EXAMPLES OF IMPROPER PACE

In the two examples below, I have purposely exaggerated the pace so that the negative effects can be spotted immediately. Both passages were paraphrased from the delightful nonsense article "The Turbo-Encabulator," circulated by the Arthur D. Little Company to its clients.

> *Pace too rapid:* The Turbo-Encabulator has a base-plate of prefabulated aluminite, surmounted by a malleable logarithmic casing in such a way that the two main spurving bearings are in a direct line with a pentametric fan constructed of six hydrocroptic marzelvanes so fitted to the ambifacient lunar waneshaft that every other conductor is connected by a nonreversible tremie pipe to the differential girdle-spring on the "up" end of the grammeters.

> *Pace too slow:* The operating point is maintained as near as possible to the h.f. rem peak. This is done by constantly fromaging the spandrels. The spandrels are of the bitumogenous type. It is interesting to note that they have a distinct speed advantage over the standard nivelsheave. Also they require no dremcock oil. This is true even when the phase detractors have remissed.

Overcrowding is the fault in the first example. Too many significant bits of information are tied together in a single sentence. The reader is not given time to sort them out and therefore cannot be receptive to new information as it comes along.

The second example has the opposite trouble. There are too many stop-and-go operations and too much drivel between significant points of information.

The next two examples were taken from actual reports. They are in no way exceptions.

> *Pace too rapid:* Since there is a 50 percent probability that the true rank is less than the median rank, the value of the median rank is obtained by integrating the marginal probability density function of the jth-order statistic of a random size n using the limits from 0 to nFj and setting this integral equal to the probability P that a random observation will have a rank at or before nFj.

Pace inefficient: The automatic controller receives all the information necessary to control the vehicle. This information is in the form of two signals generated by separate oscillators aboard the vehicle. These oscillators operate continuously at two different audio frequencies.

The first of these excerpts also shows how overcrowding a sentence will produce an improper pace. The density of information is too high for rapid comprehension — even by a mathematician. The single sentence needs to be broken into two parts: one to present the general premise; the other, the details. This simple operation would give readers enough time to organize their thoughts. The pace in the second example is not so damaging, but if it were to continue at the same rate for much longer the reader would soon tire. Consolidation of the three sentences into one would serve the reader better because it would link the key words to each other more smoothly and rapidly. One such revision might be the following, which says the same thing in half as many words.

The automatic controller receives its information from two different audiofrequency signals, transmitted continuously from separate oscillators aboard the vehicle.

But even a single short sentence can be dangerous if it contains much quantitative data. Watch what happens in the next example. No reader can easily digest more than two or three precise numbers without being permitted to pause, especially if these numbers qualify different units of measure or describe different characteristics.

Overcrowded with numerical data: Experiments showed that the camera had an error of 2 percent of the sampling rate (60 cps) for a 500-watt 13-by-23-degree sealed-beam tungsten-lamp target at 3000 feet.

Pace adjusted: Experiments showed that the camera had an error rate of 2 percent of the 60-cps sampling rate. The target was a 500-watt, 13-by-23-degree, sealed-beam tungsten lamp, located 3000 feet from the camera.

The final example illustrates pace that is too slow even for an instruction manual.

Pace too slow: A handwheel is provided for manual operation of the rotary table. This handwheel is the self-releasing type. When it is engaged, it actuates a switch. This switch disconnects the power clutch in the drive.

Pace adjusted: A self-releasing handwheel is provided for manual operation of the rotary table. When engaged, the handwheel actuates a switch that disconnects the power clutch in the drive.

In the original version, the many full stops and the repetition of key words slow the pace. But deceleration is not necessary; the sentences do not contain any numerical data, the text is not part of a set of instructions, nor are there any unfamiliar technical words or complicated ideas. The simple adjustment shown in the revision allows the reader to move along smoothly and swiftly.

FACTORS THAT GOVERN PACE

You have often heard speakers say, "Now for a change of pace . . ." What they really meant was that they were about to change the subject from something serious to something light, from something involved to something simple.

Novelists and dramatists, too, change the pace by changing the subject. They can shift from plot to plot at will — from melodrama to satire, from tragedy to comic relief.

When they change the subject, they may also change their technique of delivery. Speakers relax and adopt a less formal attitude and tone of voice as they shift to something light. Novelists quickly change from exposition to narration to dialogue — to whatever form will best fit the mood of the moment.

Most of these factors that govern pace are not open to technical writers. They must be extremely careful about introducing a personal anecdote or humorous aside. They must be formal and polite at all times. Snappy dialogue is out, gesturing is out, suspense is out, excitement is out! What's left? You'd be surprised. Technical writers can:

- Change from a statement to a question (as in this opening)
- Vary the length of sentences and paragraphs
- Emphasize important material by placing it in a prominent position or prominent construction
- Deemphasize secondary material by relegating it to a secondary position or secondary construction
- Show parallel thoughts by placing them in parallel constructions
- Relieve difficult text by introducing visual aids
- Use analogy
- Shift from straight lines of prose to columnar listing

- Break up large units of text by inserting headings and subheadings
- Change the type size or face
- Use white space to relieve the eye or isolate the text
- Modulate the "voice" by underlining, italicizing, or by placing remarks between parentheses
- Repeat, and repeat, and repeat
- Vary the choice of words

In short, technical writers can adjust the pace through effective control of the most routine elements: sentence structure, paragraphing, punctuation, format, organization, graphic illustration, and word choice.

DETERMINING THE BASIC PACE

Different types of writing — even different types of technical writing — have different basic paces that best suit the purpose and style of the type. Narration is developed historically. Time is the base and a time sequence is easy to follow. There is no trick to the organization; the characteristic pace therefore is rapid. Technical description, on the other hand, may follow any one of a variety of developments and the character of the pace will vary accordingly.

Several types of technical writing and the basic pace normally associated with each are given in the following table.

BASIC PACE IN DIFFERENT TYPES OF TECHNICAL WRITING

Type	*Basic pace*	*Reason*
Letter	Rapid	Personal; subject limited
Interoffice memo	Rapid	Informal; subject limited
Abstract	Slow	Facts all primary; crowded
Instruction manual	Slow	Time needed between statements
Historical survey	Rapid	Time as base
Technical report	Slow to medium	Introduction rapid; technical descriptions slow
Mathematical analysis	Slow	Many separate steps

ADJUSTMENT OF THE BASIC PACE

A writer can determine the proper *basic* pace by looking at the type and overall method of development of the communication he or she is about to produce. But all this does is to establish the right "ballpark." There is still much to do to meet the reader's needs: the writer must decide where in the text the basic pace needs to be adjusted and how to make that adjustment.

The following guide may be used to determine the proper pace for the individual prose units within a piece of technical writing — the sections (or subsections) and the paragraphs. Pace is not a constant, as I have already pointed out; it must vary when the complexity of the subject matter varies and when the knowledge the reader brings to the subject varies. This guide is based on the answers to two yes-or-no questions:

1. Is the reader unfamiliar with the general area of the subject?
2. Is the subject matter of the prose unit in question complex (detailed)?

In this guide, the words "normal pace" in the third column refer to the in-between or medium pace most people use in their daily writing.

GUIDE FOR CONTROL OF PACE

	Subject area unfamiliar?	*Subject complex?*	*Adjustment of pace*
Condition 1	Yes	Yes	Begin at slow pace; maintain slow pace
Condition 2	Yes	No	Begin at slow pace; accelerate to normal pace
Condition 3	No	Yes	Begin at normal pace; decelerate to slow pace
Condition 4	No	No	Begin at rapid (or normal) pace; maintain rapid pace

The technique is illustrated in the following examples. Suppose an electronics engineer were writing to a colleague familiar with the general area of integrated circuits. The subject matter contains many technical details, so Condition 3 exists.

EXAMPLE 1: PACE NOT ADJUSTED TO CONDITION 3 (FAMILIAR BUT COMPLEX)

The Personal Computer Division has just developed a new tone-decoder for use in identifying legitimate users of VHF repeaters. The compact device features an 18-pin silicon chip capable of detecting 16 pairs of tone signals and a TV color-burst crystal for timing, does not require any band-splitting filters, has good signal separation, operates on an internal 5-volt d.c. supply, and may be used in tandem, if desired, to double the capability.

EXAMPLE 2: PACE ADJUSTED TO CONDITION 3

The Personal Computer Division has just developed a new tone-decoder for use in identifying legitimate users of VHF repeaters.

The compact device features an 18-pin silicon chip capable of detecting 16 different pairs of tone signals. Other components include a TV color-burst crystal for timing and an internal 5-volt d.c. supply. Signal separation is good, without the use of band-splitting filters, and handling capacity can be doubled by using two devices in tandem.

In the first example, the pace of the first sentence is good, but the pace of the second sentence is too rapid — even for a colleague. Too many details are crowded together, separated only by commas.

In the second example, these details are spread over three sentences, with but one sentence containing two pieces of numerical data. Separation is especially important when the numbers are precise numbers and do not qualify the same unit of measure. Readers usually can digest only one number per sentence unless they are willing to slow down or reread.

THE MECHANICS OF ADJUSTING THE PACE

In the examples at the beginning of the chapter you saw how overcrowding a sentence produced too rapid a pace and how dividing the sentence into several sentences decelerated the pace. You also saw how a series of short sentences produced too slow a pace and how combining certain sentences accelerated it. You will have to call on other techniques, however, as you progress from paragraph to paragraph and from section to section in a report. I have already suggested what these are, but believe that further details on some of the points will help.

Format. "Rest areas" are essential to the well-being of anyone who has to read scientific and engineering reports. These areas customarily are found in two locations: between chapters and between sections. The space between chapters offers the longer break of the two, but the higher frequency rate of the sectional space makes it more important in controlling pace. Unfortunately, this is the space that many writers fail to provide in ample supply.

Headings are the labels for chapters and sections. These handy devices serve a multiple role: they signal the change of thought, they organize the material they govern, they permit the reader to skim, and they act as reference points. The bold centered heading looming above the text has the greatest holding power; the marginal heading, the next; and the indented heading in line with the text, the next.

White space, particularly the amount given to margins and to the distance between lines of print, has a powerful effect on pace. The eye tires rapidly when a dark block of type dominates page after page, and readers frequently lose their place when the lines are spaced too closely.

Unless they are absolutely necessary, footnotes should be avoided. They interrupt the reader unmercifully.

Paragraphs. As we have seen in Chapter 4, paragraphs are the basic building blocks of technical writing. Watch paragraphs over half a page long; they could cause trouble with pace. Too heavy a concentration of technical information is exhausting; readers need relief from paragraph overcrowding just as they do from sentence overcrowding. A series of overcrowded paragraphs is deadly — especially if some of the sentences they contain are also overcrowded.

Remember, too, that a short paragraph by itself may be desirable, but that a series of short paragraphs may produce the same undesirable effect as a series of short sentences. It all depends on the length of the series and on the function of the communication.

Most people achieve variety in paragraph length automatically, without devoting special attention to it. So don't spend your time counting words during revision of the draft. Instead, check just for the long series. Six or more short paragraphs in a row or four or more long ones should be challenged. But if they sound all right when read aloud, let them stand.

Sentence structure. Standard sentence order — subject, verb, object — is the easiest to follow and therefore the most adaptable to a rapid pace. This is an added reason for using it frequently.

A short sentence (ten words or fewer) is excellent at the beginning of a passage that will of necessity become involved. It prevents you, the writer, from engulfing the reader on first contact. It is also excellent at the end of a long dissertation, particularly if it summarizes the information in a nutshell.

Don't overlook the power (and the danger) of stating an idea as a question. The device can help the pace and sharpen your style if it is not repeated too often. Also, be sure that you answer any questions you ask.

Wording. Often the space between the end of one sentence and the beginning of the next does not give the reader enough time to understand what has been said. He or she then is not receptive to new information when it comes along.

We could give the reader more time by stretching the physical space — and this we often do by beginning a new paragraph. But some ideas are so closely related that they need to be located side by side. In these cases we have to use another delaying tactic.

I call the technique *strategic use of service words*. These words do not in themselves represent objects or ideas, but show relationship. Transitional words, such as the conjunctions, are service words. So are introductory phrases such as "on the other hand," "in summary," and "in contrast."

To pace our writing so that the reader is ready for an important point, we have to add to the words necessary to express our thoughts those words that are necessary *to convey* them. By including service words ahead of our significant words, we often can stretch the thought period just enough to allow the reader's mind to catch up with the eye.

And finally: I want to return to the author who opened this chapter, Reginald Kapp. In his book, *The Presentation of Technical Information*, Professor Kapp made this further observation:

> Most of those who have technical information or philosophical argument to present have done little towards solving [the reader's] problems of pace and timing. One hardly ever hears them discuss these problems. I doubt if they know that there are any. So low has the standard of literary craftsmanship sunk in the scientific and philosophical worlds.

Professor Kapp was not an outsider looking in. At the time he wrote his book he was dean of the faculty of engineering at the University of London and, before that, Pender Professor of Electrical Engineering, University College, London.

Part II

Special Concerns

- *Titles and abstracts*
- *Introductions and proposals*

Chapter 8

The Tenuous Title

It is surprisingly easy to acquire the usual tricks of poor writing. . . . If the proposed title, for example, means something to you, stop right there, think no further. If the title baffles the reader, you have won the first round.

"The Principles of Poor Writing"
Paul W. Merrill

In creative writing, authors often use titles to amuse, to challenge, to puzzle, and even to fool their audiences. But the authors of technical reports cannot play games with their readers, who want to be informed and expect to learn something from the first input, the report's title.

There are also practical reasons why technical writing must have informative titles. Titles play an important role in the processes of storing and retrieving information; in addition, they help a reader decide how much more attention should rightly be given to one communication than to another.

The criteria many writers and editors use to evaluate the effectiveness of a title are:

1. Does the title accurately represent the subject?
   (*Is it correct?*)
2. Are the limits of coverage stated (or implied)?
   (*Is it complete?*)
3. Is the language of the title meaningful to the intended audience?
   (*Is it comprehensible?*)
4. Has the title been expressed as efficiently as possible?
   (*Is it concise?*)

Only a few titles meet all the criteria; the majority miss at least one. Much of the time the violation is not justifiable.

DEADWOOD IN TITLES

Length is no guarantee of precision or clarity in a title. Many of us have the habit of throwing in familiar phrases that add to the word count but not to the information content of our titles. All too often, we follow a common weakness of many professional people: we use a worthless convention simply because it is established in the literature and sounds impressive.

These are some of the overworked phrases that regularly appear at the beginning of titles in technical writing:

A Report of . . .
A Study of . . .
An Investigation of . . .
An Analysis of . . .
A Discussion of . . .
A Consideration of . . .

Note that all these expressions specify in general terms only. They are bad on three counts: they occupy a strong position in the title, yet they are of no use in a literature search; they usually state the obvious; and they invariably lead to further wordiness and vagueness. "A report on" might logically be followed by some such phrase as "the use of" (both of which are unnecessary):

> *A Report on the Use of* Water as the Liquid Propellant in Project Transport

Or the opening "an investigation of" could easily tempt a writer to follow with "the effects of," which in turn introduces "using":

> *An Investigation of the Effects of Using* Water as the Liquid Propellant in Project Transport

These examples only slightly exaggerate the danger. Once a writer is trapped by the opening words, he or she inevitably begins to run off in all directions, piling phrase upon phrase. The title below is from an actual report; for obvious reasons the author will remain anonymous.

> The Results of a Study of Investigating the Effects of Using Stereographic Analysis in the Determination of the Lattice Relationships in an Iron-Nickel Alloy

You can appreciably reduce deadwood in your titles by challenging any phrase that describes what is strictly a reporting or information-gathering function.

Borderline Cases

Phrases that sharpen or refine the basic words in a title may often help the reader and should therefore not always be discarded. Test the title with and without the qualifier. If the title makes sense alone, keep it that way. Consider these examples:

Title: A SURVEY OF REPORT-WRITING METHODS IN INDUSTRY

Comment: "A survey of" is not necessary to the meaning; the rest of the title implies it.

Title: A PRELIMINARY REPORT ON PROJECT BLUNDERSTONE

Comment: "Preliminary" is informative, but "report" is redundant, because the report itself is there. Handle the phrase either as a separate label designating a type of report (the way "Progress Report" is often handled) or place the phrase at the end of the title.

Revision: PROJECT BLUNDERSTONE: A PRELIMINARY REPORT

Title: AN INTRODUCTION TO INDUSTRIAL DYNAMICS

Comment: "An introduction to" is justifiable in a title, especially in a book title. It announces that the text is for the beginner. But it would be wise to subordinate these words as a "pretitle" by setting them in a smaller type face and keeping them on a separate line.

Revision: An Introduction to (pretitle)
INDUSTRIAL DYNAMICS (main title)

VAGUENESS IN TITLES

Some writers are vague in their titles without knowing it; others know they are vague but for one reason or another don't do anything about it. Since vagueness obscures meaning, in a title as elsewhere, no author can have a strong case for advocating it.

Examine the following titles. Is the vagueness justified?

Title: A METHOD FOR MEASURING POROSITY COEFFICIENTS

Title: A SYSTEM FOR IMPROVING COMPUTER RELIABILITY

Comment: "A method" and "a system" are both vague. The titles describe in a general way what the reports are about, but specific words naming the method and naming the system (particularly the principles involved) would make these titles much more meaningful to the reader.

New Titles: MEASUREMENT OF POROSITY COEFFICIENTS BY THE SHALLOW-BORE METHOD

PREVENTIVE MAINTENANCE: AN AID TO COMPUTER RELIABILITY

We can all learn something about improving the wording of our titles from a good newspaper editor. In checking the leads and headlines to stories, editors make sure that the wording is keyed to the theme, and would never accept, for example, the vague expression "the effect of." Yet this expression appears day after day in titles of engineering and scientific writing. Behold:

Title: THE EFFECT OF ROLLING UNSTABLE AUSTENITIC STEEL AT 300°C

Comment: Not only newspaper editors but any reader might well ask "What effect?" I see no reason why words that describe the effect could not be used in the title.

One reason for the popularity of "the effect of" (and synonymous phrases) is that a precise substitute usually requires the services of a verb — and many writers feel that verbs are too outspoken in titles. Consider these examples:

Title: THE EFFECT OF INCREASING THE INPUT VOLTAGE ON THE OPERATION OF THE OLSEN ELECTRONIC SWITCH

This title does not even imply what "the effect" is; most technical people, however, would gasp at the idea of changing to:

New Title: INCREASING THE INPUT VOLTAGE IMPROVES THE OPERATION OF THE OLSEN ELECTRONIC SWITCH

Does the use of the verb sound odd to you? Perhaps you can see the logic of it in a more familiar subject:

Standard Title: THE EFFECT OF THE USE OF FERTILIZER ON LAWNS

Rejuvenated Title: FERTILIZER HELPS LAWNS GROW

Would you not agree that if we can reflect the theme of our writing in the title, without introducing gee-whiz terms or inaccuracies, we should do so? Verbs are not always necessary. A noun synonym frequently will work, as shown in the earlier example:

PREVENTIVE MAINTENANCE: AN AID TO COMPUTER RELIABILITY

SINGLE-WORD TITLES

Single-word titles are satisfactory only for pieces of writing that develop broad or general coverage of a topic. They are frequently seen in magazine and journal articles and often are very effective in drawing an audience. They should not be used in formal technical writing, however, if the subject is limited to a particular segment of a larger topic.

Title: LASERS

Comment: All types? The operation of? The design of? The uses of? The title is apt if all aspects are covered, even superficially, but otherwise it is too vague.

Title: DESIGN CHARACTERISTICS OF THE X-100 AIRCRAFT

Comment: Shortened to "Characteristics of the X-100" or simply to "The X-100," the title would not be as meaningful.

The rule of thumb is that the more restricted the area of the investigation, the more words will be required to describe it adequately in the title.

TWO-PART TITLES: THE WISE COMPROMISE

If you find that spelling out a topic produces a long, cumbersome title and you are sure it is not crowded with worthless phrases that could be eliminated, you might try the option of a two-part title. The words form a continuous title line separated in the middle by a colon. Ahead of the colon is the main idea; following the colon is the qualifying idea. The new arrangement places the key words first, giving emphasis to the general subject; at the same time, it isolates the qualifiers and thus points out their special significance. In a way, it's like having your cake and eating it too.

Examples

Title: HOW REACTION MOTORS WORK AND SOME EXPERIMENTS WITH THEM

Comment: This title is awkward and weak; elements interfere with each other.

Revision: REACTION MOTORS: HOW THEY WORK AND SOME EXPERIMENTS WITH THEM

Comment: The key words appear first (helpful certainly to the person who must catalog the report); the wording after the colon qualifies and does not interfere.

Title: THE PURPOSE AND ORGANIZATION OF THE BOSTON PLANNING BOARD

Revision: THE BOSTON PLANNING BOARD: ITS PURPOSE AND ORGANIZATION

Comment: In the original, the key words appear in seventh, eighth, and ninth places, respectively. This is not a serious loss in emphasis, but illustrates what would happen if a longer series of qualifiers preceded them.

Titles also can be split physically into two parts, known as the main title and the subtitle. The main title is usually printed in boldface type or caps and is located top center on the page; the subtitle appears below in a less-eye-catching format. This convention, which is found regularly in magazines and journals, is also perfectly acceptable for technical reports.

Examples

Main Title: WHAT SALT DOES IN YOUR BODY

Subtitle: Salt helps keep the blood neutral, distribute water, and enable the muscles to function properly.

Main Title: HEAT TREATING CAST GOLD ALLOY

Subtitle: Heating to 450°F produces optimum hardness and uniformity.

HEADINGS AS SUBTITLES

The headings most frequently found in formal technical writing serve as labels for the standard divisions of a report. The following are typical:

Introduction
Method and Equipment
Tests
Results
Discussion of Results
Conclusion and Recommendations
Appendix

Without these organizational headings the reader would be lost. But it is often possible to increase their function by changing to words that describe the content more specifically. Instead of "Introduction," for instance, you might use "History of Project Asan"; instead of "Method and Equipment," you might say "Computer Analysis of Wind Tunnel." If you handle your headings as subtitles, *they will inform as they organize*. Try it.

Examples from the Experts

Finally, here are some titles that demonstrate the rules. There really are many excellent ones in our technical literature.

"Proof That the Earth Moves"	(Galileo)
"The Orderly Universe"	(Moulton)
"Blubber into Oil"	(Melville)
"Human Ecology: A Problem in Synthesis"	(Sears)
"Digital Information Storage in Three Dimensions, Using Magnetic Cores"	(Forrester)

And on the lighter side:

"The Last Shall Be First"	(Describing automation in the shoe industry; from Arthur D. Little's *Industrial Bulletin*)

Chapter 9

The Inadequate Abstract

Mankind is learning things so fast that it's a problem how to store information so it can be found when needed. Not finding it costs the U.S. over $1 billion a year.

How to Cope with Information
Francis Bello

The problem of information retrieval has become even more acute since Francis Bello cited the $1 billion figure. The high-technology industries in particular have seen the amount triple in their research and development (R&D) facilities.

One would think that the tremendous growth in the development and use of sophisticated computers might have solved the problem. But alas, the same computers that have helped disseminate more information also have enabled their human masters to produce more.

Forecasts in the computer literature of when *completely* automated abstracting will become commonplace range from a decade to never. In the meantime, what to do? The abstracts that humans write still do not meet user needs and many are so poorly written that they are of little value. I hope that the suggestions in this chapter and the help of a friendly word processor will encourage writers to produce better abstracts faster and more efficiently. Then, at least things will improve locally.

ABSTRACT OR SUMMARY?

Many abstracts are inadequate because the authors are not sure what the design and operational specifications for an abstract are. Inspect any number of references on technical writing and you will find that authorities are not consistent in describing what an abstract is, what a summary is, and whether the terms are synonymous or whether they designate entirely different elements of a report and should therefore be used selectively.

Since the confusion results from a problem in semantics and from a personal choice of terms, I will use only "abstract." An abstract appears apart from, and ahead of, the text. It is more of a "sampling" device than a review device. In conjunction with the title, it tells the readers what the main thoughts of the communication are so they can decide whether or not they want (or need) to read the details that follow.

TYPES OF ABSTRACTS

The abstract performs its service to the reader in one of two ways:

1. It acts as a *report in miniature*, a capsule version of the main report, highlighting the main points.
2. It acts as a *prose table of contents*, indicating the main topics that are covered in the body.

Some textbooks refer to the report-in-miniature type as "informative abstracts" and the table-of-contents type as "descriptive abstracts." These terms, however, also confuse many writers, since the informative abstract contains technical descriptions and the descriptive abstract can be said to inform. Perhaps some of the confusion can be avoided by substituting "indicative" for "descriptive"; the table-of-contents type "indicates" what the report contains.

The method characteristic of each type can be illustrated by comparing two abstracts of the same article.

Informative Abstract

QUIP: A NEW PACKAGE FOR IC'S

The XYZ Company has developed QUIP, a 64-lead package for integrated circuits that uses two rows of pins along the sides instead of the customary one row. This enables an IC to be packaged in a case only 1 5/8 inches long, thus increasing the strength and rigidity. It also enables the internal conductors to be shorter, resulting in an estimated 10 percent lower pin-to-pin capacitance, lead resistance, and inductance. The top of the case can be easily removed with a screwdriver so that probes may be inserted to check the circuitry while the IC is functioning.

Indicative Abstract

The XYZ Company has developed a new package for ICs, called QUIP, that uses double rows of pins to accommodate more circuitry in less space than is possible with single rows. The main design features and the test results are given in this article.

Other distinguishing features of the two common types also are:

- For a given piece of writing, an informative abstract will run slightly longer than an indicative one.
- An informative abstract follows the style and language of the original; an indicative abstract uses passive verbs such as "are described," "are given," and "are discussed."
- An informative abstract often contains much quantitative information; an indicative one does not.

Which Type of Abstract Is Preferred?

Neither type of abstract is recommended for *all* communications. Make your selection first by analyzing the nature of the subject matter and then by establishing the objective of the reader.

An indicative abstract can be written from almost any type of communication; an informative abstract, on the other hand, is not that flexible. For example, it is not suitable for a textbook or for any long and involved publication. There would just be too many significant points to abstract. Nor is it satisfactory for covering material in which understanding each point of text depends on understanding the point that preceded it (such as in the development of a theory or in an elaborate set of instructions). The informative abstract is at its best with shorter technical communications (report or article length) in which the reduction of text from original to abstract does not have to be so great. Reports on tests and experimental investigations lend themselves nicely to the informative abstract, as do reports that answer "how much?" or "how many?"

If the nature of the subject matter rules out the informative type of abstract, then use the indicative. However, if the subject matter does not eliminate the informative, determine which type will better satisfy the needs of the reader.

The informative abstract is preferred by readers who wish to get the main points (such as results, conclusions, and recommendations) *without* reading the report itself; by those who must take action on the main points immediately but who will eventually read the report; and by those who wish to know special technical details without having to commit themselves to reading the full report.

The indicative abstract is preferred by readers who wish to know what the general coverage of the writing is, what the subdivisions are, and how the material is developed. Perhaps I should say "acceptable to" rather than "preferred by" these readers, because an informative abstract might satisfy them even more if it provided coverage of contents while it reported the main facts.

The indicative abstract is a general-purpose device, so to speak, and should not be used when a special-purpose device will do a better job. *Failure of the writer to make this distinction is the main cause of inadequate abstracts.*

PROCEDURE FOR SELECTING THE PROPER TYPE OF ABSTRACT

Step 1. Determine the nature of the communication.

Step 2. Determine whether an informative abstract is ruled out.

Step 3. Determine reader preference.

Let's assume that the communication is a technical paper on computer hardware development, addressed to designers of computer memories.

Step 1. The paper is fifteen manuscript pages, contains quantitative information, and presents specific results and conclusions. Although theory is outlined, the emphasis is on technical details.

Step 2. An informative abstract is not ruled out by the length of this paper (fifteen pages) nor by the treatment of subject matter.

Step 3. Since both types of abstract are eligible, choice should be based on which one the reader would prefer. And since the intended readers are specialists in the subject area being reported, we know that they would be interested in details and in getting quantitative information as rapidly as possible. These facts make selection of the informative abstract mandatory.

The title and informative abstract might read:

USING ULTRAFAST LIGHT PULSES TO PROCESS AND STORE INFORMATION

The XYZ Computer Laboratory will soon start the groundwork to develop a prototype ultrafast memory that will utilize extremely short light pulses as the storage vehicle. The pulses are produced by laser and travel at 186,000 miles per second. Each lasts for only

16 femtoseconds (16×10^{-15}) and covers just two ten-thousandth of an inch. The underlying theory relating to storage of information is that an electronic material can be made to alter its behavior if pumped with energy from these pulses.

Let's now assume that the communication is an article addressed to management readers and that it discusses the state of the art of methods for generating high temperatures.

Step 1. The article is general theory, but not highly analytical. It runs about twelve manuscript pages. The purpose is to urge managers of R&D organizations to encourage their people to develop industrial applications, utilizing the knowledge from this basic research.

Step 2. The length does not rule out an informative abstract, but the scope and general treatment of the subject suggests an indicative approach.

Step 3. Management readers usually prefer an indicative type of abstract when the communication involves detailed technical information. This article presents quantitative details on two methods that probably would be meaningless to them in an abstract. A straight informative type, therefore, is ruled out. However, a straight indicative abstract can be dull to the point of not attracting interest. So whenever it is possible to identify specific items by describing them on a general level, by all means do it.

A straight indicative abstract might read:

GENERATING HIGH TEMPERATURES FOR INDUSTRIAL USE

This article describes in detail two methods for generating temperatures above the melting point of tungsten (10,000°C). The procedures for employing these methods are described and the problems associated with each are discussed. Possible applications by industry, using these methods, are then presented.

An indicative abstract, flavored by touches of the informative type, might read:

This article presents two methods for generating temperatures above the melting point of tungsten (10,000°C): the solar furnace and levitation. Although not as practical as the solar furnace, levitation — heating a metal suspended by an electromagnetic field — has long-range promise. Both methods could be useful in such industrial applications as coal gasification, nitrogen fixation, and production of new heat-resistant alloys for space vehicles.

In brief, write a straight indicative abstract only if the length and treatment of subject matter rules out the informative type. Whenever possible, write either a straight informative type or a combination type, depending on the needs of the intended reader.

WHAT ABOUT LENGTH?

Everyone agrees that an abstract should be short. But how short? Some company style manuals suggest a maximum of one page of abstract for every thirty pages of text. This is not an unreasonable ratio, and serves as a useful yardstick. However, if followed religiously it often encourages inefficiency.

Instead of planning by ratio, you should make every effort to keep *all* abstracts from running over one-half page or 150 words. Naturally, a few may need to be longer if the informational needs of the reader are to be met. But the majority will easily fit within the 150-word limit. In fact, many will not need to be longer than 75 words.

Extremely short abstracts (one or two sentences) raise questions of justification. If the text is short, is an abstract necessary? Can't the reader get the information just as easily by skimming the text? Probably, but we still need the abstract to assist in information retrieval.

If an abstract merely repeats information already given by the title, should it be retained? No. It should be rewritten. Consider the title and abstract as forming a communication unit and behaving much as a paragraph. The title, in effect, announces the topic; the abstract develops the topic. The whole has unity, transition, and movement; any redundancy that remains is purposeful.

The convention of confining abstracts to one paragraph often leads to extreme overcrowding. If the material will be easier to read in two or three short paragraphs rather than one long one, there is no good reason for observing the convention.

SUGGESTIONS ON LANGUAGE

All abstracts have a high density of significant words, but all need a few noninformational or "service" words to make the reading easier. Don't forget to supply transitions, especially after taking whole sentences bodily from the text.

The time to establish the key words of your communication for the reader is in the title and the abstract. These words are the nouns and verbs that name, define, and describe the important ideas in your writing. They are the words that would permit accurate filing, referencing, and retrieval of your report. Use standard terms whenever possible and use the same term in the abstract that appears in the body. All of the professional technical societies have published lists of terminology they consider standard for their fields. Be sure to have up-to-date copies of these lists on hand.

All but the most familiar abbreviations and acronyms need to be spelled out at first appearance. NASA, for example, is sufficiently well known to stand by itself; DCC (Document Control Center) is not. In any event, avoid extremes. This acronym goes too far: EGADS (Electronic Ground Automatic Destruct Sequence)!

It is safe to assume that the abstract of a report or paper will never appear without the main title. Awkward redundancy between the title and the first sentence of the abstract should be avoided. For example:

Title: DIGITAL SIMULATION OF RANDOM VIBRATIONS

First sentence of abstract: This thesis describes an investigation of digital simulation of random vibrations.

The first six words of this sentence are worthless; the remaining five do not contribute any new information. The needless redundancy would never have occurred had the writer thought of the title appearing directly above the abstract (as it does on an abstract card).

MECHANICAL AIDS

There is an unwritten convention within the technical community that warns writers not to use graphic illustrations in abstracts. Undoubtedly this came about because of limitations on the space assigned to abstracts and also because many abstracts are published separately, some on abstract cards. But is it wise to rule out illustrations entirely? They certainly could be used in reports that do not go beyond the family, so to speak, provided they are kept simple and provided they do not make the abstract occupy more than one page. If you believe that a sketch or diagram would make your next abstract clearer (and if you know you won't be breaking any house rules), go ahead and try one.

Subheadings can also be used in abstracts more than they are. There certainly is no law forbidding their use in abstracts. Granted, they may be overdone and they would not be in order in the majority of cases. But they too should not be ruled out categorically; the abstract of the moment is always the important abstract, and it just might benefit from the improvement of organization that a few subheadings would provide. (Remember, of course, that you cannot often use a lone subhead. You must have two at least.)

THOUGHTS ABOUT COVERAGE

Abstracts should be written *after* the main body of the communication, not before. The original communication will thus shape its own image, and not vice versa.

The abstract should never contain information not presented in the body of the report.

A well-written, informative abstract is a replica, in miniature, of the original. It, too, has a beginning, a middle, and an end, with emphasis on the key ideas and/or results.

Most writers outline their material before they write the first draft. The enlightened ones also prepare a statement of thesis, which they use as a guide for selecting and rating the raw material they have collected during the investigation. This statement of thesis relates the initial problem and objective to the subsequent results and conclusion. It prescribes the course of development the author wishes his or her message to take in order to communicate successfully with a specific audience. Reduced to its key words, the statement of thesis will form a meaningful title; expanded with supporting details, it will form a meaningful abstract.

OVERVIEW VERSUS ABSTRACT

Many management executives who do not have a science or engineering background prefer what they call an "overview" to the customary abstract or summary. As technical writers and editors may well be called upon to write overviews, a few suggestions might help.

While working on revising this chapter, I asked a management friend to compare an overview with an abstract. This is what he said: "Think of an overview as a synopsis. As such it usually follows a narrative style rather than an expository one. It usually is longer than an abstract. For example, abstracts of journal articles normally run no more than a single paragraph; overviews of company reports of comparable length normally run a page or two. The type of abstract you call 'informative' is closer to an overview than an indicative type is. However, an overview may well include an evaluation or a recommendation that is not in the text itself. In that respect, it is more like a letter of transmittal than an abstract."

I then asked my friend to say in one word what the most important aspect of an overview is. His reply: "Readability." But then he added: "Don't think for a minute that we're not looking for accurate, concise reporting. It's just that we don't want to get messed up by a lot of technical jargon; we've got enough problems with our own gobbledygook!"

Chapter 10

The Improper Introduction

> ***Trying to explain a new scientific theory without first introducing its general character and purpose is like trying to entice a woman to make love by winking at her in the dark: the intent never is realized.***
>
> Norbert Wiener
> (in a conversation with the author)

In questioning several groups of college students recently on whether they were having difficulty writing their B.S. theses, I was not surprised to find that about a third reported trouble with the introduction. This figure is not a general statistic, of course, but it does reflect a problem that often arises whenever engineers and scientists have major writing jobs to do.

The main reasons the students gave for having trouble with the introduction were:

1. I don't know exactly what an introduction should do for the reader.
2. I don't know how to determine what the reader already knows.
3. I don't know how to organize the material effectively.

The order in which these reasons are listed is not intended to indicate a priority; one cause of trouble can be as damaging as the next. Let us briefly consider each of these stumbling blocks.

THE INTRODUCTION AND THE INTENDED READER

A quick way to understand what an introduction should do is to examine it in relation to the communication as a whole. The introductory section supplies preliminary material to help the reader quickly understand and appreciate the real message that the communication carries. Its role is to bridge whatever gap may exist between the writer and the audience so that the purpose of the communication can be fulfilled.

The gap is primarily one of information. For example, if a reader is to understand why a certain procedure was followed in a research project, he or she will need to be briefed on the problem that brought about the project. A proper introduction would provide this information, no more but certainly no less.

The gap may also be one of motivation. The reader may need to be prodded, challenged, or otherwise stimulated to read. When this is the case, the introduction acts in the same way as the lead in a magazine article. Working closely with the title, it builds up reader interest in the "story." If it is well written, it does this simply and honestly, never promising anything the author can't deliver.

In short, an introduction in technical writing is designed to prepare the intended reader intellectually and emotionally for the serious job of interpreting and evaluating a message.

DETERMINING WHAT THE READER ALREADY KNOWS

If the intended reader (i.e., the primary audience, whether one person or many) has a technical background similar to yours, he or she will need at least as much briefing about the assignment as you did. Remember, you are at the end of the investigation; your reader is at the beginning. Think back to the time you were assigned the work. What briefing were you given? What questions did you ask? What preliminary investigating did you have to do? What information did you discover later on that would have helped you more if you had found it earlier? Your own needs can serve as a reliable gauge.

If the intended reader has a technical background in a specialization other than yours, you will have to carry your questioning a step further. What explanatory information in addition to that required by the first audience will be needed? This question has no pat answer, but most of the time the supplementary material involves definition. The common ground of writer and reader is now restricted to their general technical knowledge. To expand this ground so that the reader will be able to understand the specialized aspects of the communication, you will need to define many terms and concepts that did not need defining for the previous audience.

As these special items appear in the introduction, relate them whenever you can to something that is already known and thus familiar to the reader. Handle your descriptions in as simple, concise, and straightforward a manner as possible. You can check on your judgment by asking someone with a background similar to the intended reader's to comment on what you have written.

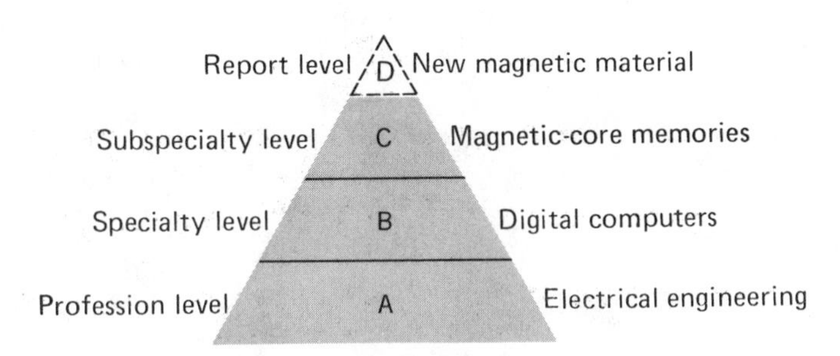

Illus. 10–1. Levels of reader context.

Illustration 10–1 outlines the process of coordinating the technical level of the subject with the background level of the reader. The electrical engineering profession is used as the guinea pig. The bottom block of the pyramid, A, represents the technical knowledge common to everyone in electrical engineering. It naturally has a broad base and includes many people. The second block, B, represents the added knowledge common to everyone in a specialty of electrical engineering, digital computers. It includes fewer people. Block C represents the highly specialized knowledge common only to those involved with magnetic-core memories. D is the area representing the subject of the report. Presumably, the information it contains is new to the readers.

Suppose that an engineer in the C group has just finished investigating the operating characteristics of a new magnetic material and is asked to report the findings. If the report is to go to magnetic-core specialists, the communication needs to be introduced only within the context of C. If, however, the report is going to computer engineers specializing in areas other than magnetic-core memories, the introduction must fall within the context of B. Should the audience be electrical engineers, but in radar work or optics communication, say, rather than computers, the introduction would have to be formed within the general context of A.

Any writer can work out a pyramid profile of the intended reader, compare it with his or her own profile to see which blocks they have in common, and then use the highest of these as the context within which to introduce the subject.

ORGANIZING THE INTRODUCTORY MATERIAL

If the introduction is to help the reader, it must follow some organizational pattern that will proceed from the familiar to the unfamiliar, from the general to the specific, from doubt to open-mindedness. One way to accomplish this development is to invert the pyramid, as in Illus. 10–2.

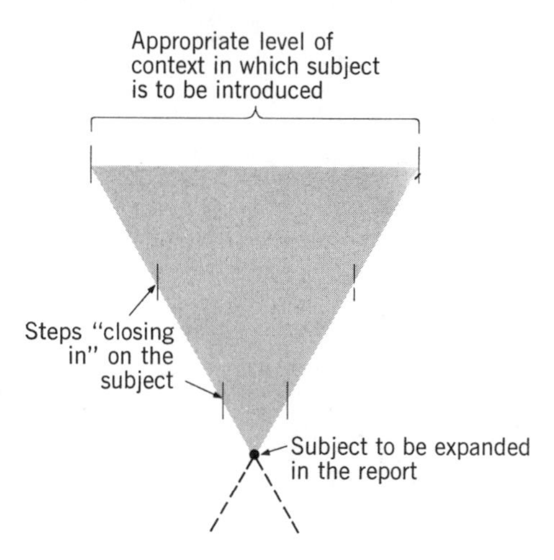

Illus. 10–2. Inverted-pyramid technique for developing the introduction.

The apex of the pyramid, now at the bottom, represents the special subject to be presented in the report. The base line above represents the broader area of context from which the reader's preliminary understanding of the subject must be drawn. The development involves closing in on the subject through one or a series of logical steps. The number of steps required depends on the length of the base line, the highest level of context in which the writer and reader meet.

A single step usually is sufficient when the writing is addressed to colleagues. One common pattern is the problem-solution sequence. The problem explains the "why"; the solution, the "what." Illustration 10–3 is an example.

More than one step may be desirable to introduce the same subject to a wider technical audience. Illustration 10–4 shows how this can be done.

THIN-FILM TRANSDUCERS

Westinghouse has developed a method of producing transducers that will vibrate at ultrahigh frequencies but not shatter under their own vibrations. The technique involves depositing a thin film of cadmium sulfide from vapor in a vacuum chamber onto a substrate . . .

Illus. 10–3. Example of a single-step introduction.

THIN-FILM TRANSDUCERS VIBRATE AT ULTRAHIGH FREQUENCIES

Piezoelectric transducers, which convert electrical pulsations into mechanical vibrations, ordinarily consist of a thin wafer of a crystalline material such as quartz. To achieve high frequencies, these ordinary piezoelectric crystals must be made so thin that they shatter under the vibrations they generate; they are so fragile, in fact, that it is nearly impossible to handle them without breakage.

A new way of making crystal wafers produces transducers that are much more durable. These transducers have been operated at frequencies up to 75,000 megahertz, and the technique should eventually provide frequencies approaching a million megahertz.

The transducers are thin films of cadmium sulfide deposited on a substrate from vapor in a vacuum chamber . . .

Illus. 10–4. Introduction expanded for a wider audience.

Another familiar pattern is the problem-purpose sequence. Here a statement of the problem exposes an underlying cause that produces a general undesirable effect; the purpose relates the immediate investigation to the cause-effect problem. Illustration 10–5 is an example of this sequence.

Some writers prefer to begin with a statement of purpose, following with the statement of problem. Their reason for the preference is that the reader is informed immediately of the objective of the investigation. The reason is valid and the pattern is acceptable. But there are some disadvantages:

- Transition from introduction to body is not as smooth.
- Motivation cannot be built up as effectively.
- Awkward redundancy sometimes exists between the title and the opening of the introduction.

These negative effects can be somewhat lessened if the impact of the purpose is held to the end by dividing the statement into two parts. The

PROPOSAL FOR INSTRUMENTING A BLIND MAN'S CANE

A cane still is the best inanimate mobility or guidance device available to the blind person. It is easier to use and is more reliable than any of the existing sonar or optical scanning and detecting devices that have been developed during the past twenty years.

It is not known why a cane works as well as it does; nor have all of the useful stimuli it provides the user been evaluated. Development of a better device, therefore, cannot be expected until the job of evaluating the cane's performance is completed.

The purpose of this investigation will be to measure all of the physical variables that the cane produces and to correlate these with the user's responses so that the relative importance of the stimuli can be established.

Illus. 10–5. Example of problem-purpose sequence (from a student's report).

opening would explain the purpose in terms of a general solution. This would be followed by a detailed statement of the problem. The close would restate the purpose in terms of the immediate objective of the investigation.

Not all introductions can be developed smoothly in one or two steps. The reader may need to be informed of many things — some historical, some technical, some political. Where the emphasis is placed will depend on the reader's background and present interest. Illustration 10–6 provides a good example of a multistep introduction that uses historical development.

The common danger of a long introduction is that readers will lose interest and quit. To lessen the chance of this happening, writers should never assemble the introduction into a single, many-page paragraph. Instead, they should see if the material might not be divided conveniently into logical parts (the steps of the pyramid). Each part might be a paragraph or, if there is a great deal to say, a group of paragraphs tied together by a heading. Readers are thus given some freedom of movement and can skim if they wish. (Another scheme, of course, is to relegate some of the details to an appendix.)

To Conclude

Readers need and want an introductory section in their reports, but don't enjoy reading any more preliminary material than is absolutely necessary. Writers can trim the introduction to fit their readers' needs by beginning their account at the proper level of context and then closing in on the subject by a series of logical steps. Some introductions have to be long because a great amount of historical background is needed to bring the reader

GROWTH OF THE TECHNICAL WRITING PROFESSION

Technical writing has a long and exciting history. Its beginning cannot be set exactly, but we do know that the exchange of written reports on matters of engineering, science, and technology was established practice in many countries before the birth of Christ. Of the early manuscripts that have survived, perhaps the most popular has been *De Architectura* by Marcus Vitruvius Pollio, chief artificer of Augustus Caesar. Compiled in 27 B.C., this work summarized all that was then known about city-planning, building materials, civil and military structures, interior decoration, hydraulics, technical instruments, and machines and engines.

Although the art of technical writing is almost as old as science itself, the profession of technical writing has a recent beginning. In this country, before World War I, the writers were the doers. Scientists and engineers, whether in industry, business, or government, wrote their own copy. Not until the industrial decentralization of the 1920s did they receive any editorial assistance from the staff level, and this unfortunately ended at the start of the Depression.

The real impetus came in World War II, when a special task force was recruited to deliver the paperwork. These "technical writers" soon established beachheads of their own in the documentary wastes of Washington, and by the end of the war had extended their front lines far into business and industry.

The object of this paper is to trace their progress since then. In the time allotted, I cannot account for every branch of so widespread an activity. But I believe I can present an honest picture of the general trends. To simplify the description, I will treat technical writing and technical editing as a single profession; however, this does not mean that I consider the two skills identical.

Illus. 10–6. Example of multistep introduction.

up to date. In these cases, length can be made less objectionable if the material is divided into sections and headings used.

If, after you've worked awhile on polishing your introduction, you still don't like the way it reads, remember that many an introduction has blossomed after the opening paragraph was lopped off!

Chapter 11

The Awkward Proposal

My students have never been taught how to write a proposal. They flounder about, become terribly annoyed, and curse me for giving them the assignment. Why don't you put a chapter on proposals in your book?

Peter Griffith
Professor of Mechanical Engineering, M.I.T.
(in a note to me on a student's paper)

This chapter concerns itself with problems that begin when someone is confronted for the first time with the task of writing a modest research proposal.

The suggestions are directed primarily to science and engineering students, undergraduate or graduate, but should also be helpful to anyone new at proposal writing. The material that will be covered can, I believe, be applied directly to a variety of required writing assignments. Some that come readily to mind are:

- A proposal for a research project in a laboratory class
- A proposal for a term project in a course
- A proposal for a project to be conducted in a cooperative program with industry
- A proposal for a grant from a foundation
- A proposal for a bachelor's or master's thesis

Over the years I have read proposals for all of these enterprises. Some of them were well written, but the majority, it is fair to say, suffered from a major fault: it was difficult to tell exactly what the proposal proposed. The cause was always a combination of factors involving both style and content, pointing to a definite need to return to instruction in the basics.

DEFINITION OF A PROPOSAL

A proposal is a formal statement from an originator to a reviewer about the possibility or probability of achieving a goal allegedly beneficial to both parties. The content of a proposal identifies the problem to be solved or the need to be met and presents a plan for achieving that end.

Proposals vary in size and purpose, from a short proposal of a term paper from a student to an instructor to a set of tomes from a contractor to the military detailing a bid for a major contract.

No matter how simple in form, good proposals are not easy to write. They are, after all, a selling device; but in the technical community at least, they must be void of the trappings of Madison Avenue. A dull, awkward proposal, full of footnotes and technical jargon, must also be avoided.

These limitations narrow the field to a specialized communication that must be both subjective and objective in its execution. Clearly, writing a good proposal is not an easy task. So to begin with, let's look at an outline of what an acceptable research proposal should contain:

GENERAL OUTLINE OF A PROPOSAL

I. Descriptive title

II. Specific statement of what is proposed

III. Background
   - (a) The situation that brought about the problem
   - (b) Why the problem needs to be solved
   - (c) What has been done to date: e.g., tests, literature search
   - (d) In general terms, how the investigation will solve the problem (this states the hypothesis)

IV. Scope
   - (a) The goal of the research in terms of the results
   - (b) The specifications set by the author or the reviewer
   - (c) The limitations imposed on the investigation
   - (d) The assumptions used in the analysis

V. Methodology
   - (a) Tasks
   - (b) Methods and procedures to be employed
   - (c) Time schedule

VI. Results
    (a) How they will be compiled
    (b) How they will be evaluated

VII. Project requirements
    (a) Staff (names, qualifications, and assignments)
    (b) Equipment needed (on hand and to purchase)
    (c) Facilities needed

VIII. Costs
    (a) Breakdown into categories
    (b) Method of payment

IX. Publications
    (a) Progress reports
    (b) Final report
    (c) Internal communications

COMMENTS ON THE ENTRIES IN THE OUTLINE

Title

This should contain the key words of the hypothesis, leading directly to the statement of proposal. Consult Chapter 8 for suggestions on forming titles.

The Statement of Proposal

If you miss here, the whole proposal will suffer. Try to state your case in a single paragraph. The "what" and the "why" of the investigation should stand out. As a professor of mine used to say, "What will be your crucial contribution?"

Background

(a) Investigations are undertaken because either a problem has to be solved or a challenge met. What brought about the problem or challenge? How did you learn of it?

(b) Why is it important to you, to your reviewer, instructor, supervisor, or to your field of study that the problem be solved?

(c) What previous work has been done to solve the problem? By others? By you? If by others, what did they accomplish? Where did they succeed? Where did they fail? Why? If you have made a previous study of the problem, state what you did and evaluate it.

(d) A proposal is based on a hypothesis. What is yours? On what do you base it? How will the research you plan solve the problem? What is your estimate of probable success? On what do you base this estimate? (It's better to be conservative in this estimate than too optimistic.)

The Scope

(a) Specifically, what main result or results do you hope to achieve?

(b) Spell out the specifications you will follow to achieve the results and any that define the results themselves.

(c) All investigations have to be conducted within certain constraints or boundaries. What are yours? Is your goal compatible with the time available? The equipment? The facilities? The budget?

(d) Your reviewer will wish to know what assumptions you have established and if these assumptions are valid. State your reasons for believing they are valid.

Methodology

(a) List the tasks in the order in which you expect to perform them. What will each task achieve? How will they fit together to enable you to reach your goal?

(b) Name your methods and procedures. Go into detail only on those that are special, new, or unfamiliar to your reviewer.

(c) Work out a bar chart, with your tasks listed vertically and plotted horizontally along the abscissa in units of time to show the beginning and the end of each task and where any tasks overlap. Be sure not to forget to list the task of writing your report!

Results

(a) How do you plan to compile and record your results? Will you use any special features — e.g., oscilloscope traces, computer visuals?

(b) The reviewer will look carefully at how you plan to evaluate your results. Be specific here (terse statements such as "I plan to check them out on the department's computer" will not be enough).

Project Requirements

In a large proposal, such as a company's bid for a contract, this is known as the "Capabilities" section, and the information presented is known as "boilerplate" — ready-made material from the public relations file for uses of this sort. Since you may have no formal "capabilities" to offer other than yourself, you will have to list your requirements and indicate how you plan to meet them. For example, what equipment and facilities will you need? How will you obtain them? Try to be realistic in your requirements. Don't pad your figures, but don't understate your requirements either. List exactly what you believe you will need in order to complete the research as you have proposed it. If you believe that you might stand no better than a 50-50 chance of obtaining one of your requirements as specified, be prepared to scale down that requirement. Work out an alternative in advance, but don't report this in your proposal.

Costs

As a student, you might not have a great deal to say here if your costs are built into your educational program. So your instructor might not ask for this section in your proposal. But it's a good idea to get some experience in costing; you'll be involved with it all too soon in your career. An excellent opportunity for learning the economic ropes is to write a proposal for a research grant from a foundation. A careful forecast of costs is vital to the success of this type of proposal. Be sure to inspect the literature on funding issued by the foundation and, if possible, find a friend who submitted a successful proposal and see how he or she handled costs.

Publications

Most research projects require a minimum of two types of reports to document the investigation: progress reports and a final report. For an undergraduate research project, usually lasting only a semester, probably one progress report midway through the project is sufficient. More would be expected for a professional project. The first progress report uses the proposal as its point of departure: the proposal sets the general course of the investigation; each progress report marks the progress along that course as set by the time schedule.

The final report is a complete story of the investigation. It is a formal document, with references and a bibliography. An outline of the main body of a typical final report is given in Illustration 11–1. Note that it gives ample details to be useful in writing a first draft.

For internal publications it would be wise to keep a log even if one is not required. You definitely would have to do so if you were working for an R&D organization. (For example, a patent suit could be thrown out because of inadequate proof of when the research was done.)

I also highly recommend that you keep a journal. It can be as informal as you like because it is a personal document. Unlike the log, it is subjective. With it you have a chance to record your guesses, disappointments, postmortems, reactions, and reasonings that otherwise you might forget. These sometimes turn out to be as important as the formal procedures and results that you make public. A journal also will help you when you write

AUTOMATIC EXTRACTION OF PITCH IN SPEECH SOUNDS

INTRODUCTION

1. Problems to be solved in designing a system for automatic recognition of speech sounds
2. Methods previously used
   (a) Sort out lowest freq. component
   (b) Detect differences in freq. between adj. freq.
   These methods do not make use of all info. in every freq. harmonic. Result not reliable

METHODS OF ANALYSIS

1. "Absolute difference" method
2. "Autocorrelation analysis of a periodic signal"
3. Both methods have advantage over conventional:
   (a) Can be applied even when freq. comp. is absent
   (b) Utilize entire harmonic component
4. Both methods investigated & compared

EXPERIMENTAL PROCEDURE

1. Test signals
   (a) Natural voices
   (b) Synthetic voices
   (c) Pitch adj. to 100,200,300,400,500 Hz.
   (d) 3 different loudness levels
2. Preparation of data
   (a) Speech sounds converted to voltage
   (b) Voltages conv. to digital
   (c) Stored as binary numbers in computer
   (d) Important parameters selected
   (e) Sample size adopted
3. Computer programs
   (a) 3 programs for parameters
   (b) Precautions re. overflow
   (c) Average times 2.4 vs 90 sec.

RESULTS AND DISCUSSION

1. Synthetic vowels
   Both methods worked without single failure
2. Natural vowels
   (a) Both methods o.k. for 200 Hz.
   (b) Deterioration at 100 Hz.
3. Existence of sizeable subharmonic compont. Auto. less affected by it

CONCLUSION

Both methods work perfectly for periodic waveforms, regardless of freq. spectra. For same accuracy, auto. correl. requires large sample & gives short-term avg. Abs. Diff. needs only small sample & gives instant pitch

Illus. 11–1. Outline of the main body of a research report.

the progress report(s) and the final report. Keeping a journal is a fine habit to acquire, one that will help you for the rest of your career.

Along the same line, you might wish to use a cassette recorder to help you gather information before the proposal is written and after it has been approved. The convenience is especially appreciated during a literature search and an interview.

BEFORE YOU BEGIN TO WRITE

In order for you to propose research on any problem, you must identify the problem and then determine if a solution is feasible. These steps usually are accomplished by conducting a preliminary investigation of the problem area. I assume that you have done all this and have decided to go ahead. Before you begin to write, however, you need to plan your strategy.

First, ask yourself again "What is the problem that needs to be solved?" Write down an answer to this question and go over it until you are satisfied that your response is correct. Once you do this you will have delineated the reviewer's primary interest. You now have the hypothesis from which you should develop the framework for your proposal.

Your preliminary findings that relate directly to the hypothesis form the primary evidence in your proposal. Those that relate indirectly are secondary. Concentrate on the primary bits first; fit in the secondary ones later on.

Next, determine what the scope of the research should be to produce the results that will enable you to restate the hypothesis as a thesis. Reexamine your preliminary findings. They should suggest what the nature (type) of your research should be (analytical, experimental, etc.). Then consider what information your reviewer will need in order to accept your evaluation of the results. These two elements (nature of the research and the reviewer's needs) govern the scope of the investigation; they answer the question "How much detail?"

THE WRITING: CONTENT AND MECHANICS

Abstract. Although the sample outline of a proposal presented earlier did not list an abstract, you should use one. Make it as informative as you can (with the title it should form a proposal in miniature). Hold it to one paragraph, no longer than half a page. Put it right after the title page.*

* In a long, multichapter proposal, a summary as well as an abstract often is included. The abstract still is kept short and represents the thesis of the proposal. The summary may run several pages, giving important details from all of the chapters.

Table of contents. You should also include a table of contents if the proposal runs over ten to twelve pages. Include all main headings and first-degree subheadings. Assemble it last so that you can insert page locations. Locate it after the abstract.

Bibliography. Undoubtedly you will need a bibliography of the literature you consulted. List by last name of author in alphabetical order. If a book, include title of work, city of publication, publisher, and date. If an article, name the article, the journal, the publisher, and the date.

Footnotes. Use reference footnotes sparingly — if at all. If you have only a few references, insert them parenthetically within the text where you need the reference, like this: (Rollins, C.J., 1983). The reader can turn to the bibliography for details.

Appendix. If you wish to include data sheets and large exhibits in your proposal, place them in an appendix. Be sure to mention in the text that you have placed them there. Number your appendices in capital roman numerals and give each one a title:

APPENDIX I DATA FROM EXPERIMENTAL TEST

Headings. Be sure to use headings that describe the content of the passages they introduce. Main headings tend to be general, which is O.K. if they describe content rather than function. Subheadings *must* be specific. Refer to Chapter 8.

THE WRITING: ATTITUDE AND TONE

In your writing, the correct attitude toward your reviewer is that he or she is intelligent and interested in the problem you wish to investigate but is not greatly informed on its details. Although you may have been assigned the work by an instructor or supervisor, you should not assume that that person is an authority on your problem. (Even if this is the case.) If you do assume this, you might hesitate to give sufficient background information in your introduction for fear of oversimplifying the case or being a bore.

Why would this be bad? For two reasons: first, the reviewer just might not be familiar with the details of the problem; second (and the more likely reason), the reviewer would be familiar with the problem but wish to be certain that *you* are. So approach the writing as though the person at the receiving end were learning about the situation for the first time. This attitude should prevail throughout the proposal as well.

The tone in your proposal should be confident but not so that it forces you into a style that isn't natural. You will serve the same purpose, actually, by being straightforward: avoid *hedge words, meaningless qualifiers,* and *all words with negative connotations.* "Hopefully," "should," "probably," and "unfortunately," for example, do not instill confidence in your work. On the other hand, do not avoid using the pronoun "I." "I believe that this solution will work" shows more confidence than "It is believed that this solution will work." Both express an opinion, but the first is more positive. (If you were writing a proposal on behalf of a company, however, the first person "I" would be out of order, even though you had conducted the investigation.)

How you present your results, conclusions, and recommendations also affects the tone. Don't bury positive points or assets by overcrowding with nonessential details. (This happens when a writer wishes to impress the reviewer with all the work he or she did.) Choose the type of visual aid that immediately displays the data you wish the reviewer to remember. This does not mean that you are to suppress information, only that you should put your best foot forward.

Part III

Aids for the Writer

- *Worksheets for outlining and writing a draft*
- *Check lists for editing and testing the writing*
- *Suggestions for using a word processor*
- *Helpful references on all phases of technical communication*

Chapter 12

Techniques and Devices

Read over your compositions, and when you meet with a passage you think is particularly fine, strike it out.
Samuel Johnson

This chapter suggests techniques that fit the conventional mode of writing and revising a manuscript; using a word processor to help accomplish these tasks is covered in Chapter 13. The worksheets reproduced below have been used successfully by undergraduate and graduate students who previously had no organized plan for attacking a writing assignment.

WORKSHEETS FOR PREPARING A TECHNICAL REPORT OR JOURNAL ARTICLE

Questions to be Answered Before You Begin to Write

1. In as specific terms as possible, what is the subject of your report or article to be?
2. Who are the intended readers? What are their backgrounds regarding the subject?
3. Why will these readers be interested in reading what you have to say?
4. What use will the readers wish to make of the information in your report or article? (Will they wish to use it in some special way?)
5. What specific questions do you believe the readers will wish to have answered concerning the report as a whole?
6. How might you tailor the structure and format of your report or article so that you would have a good chance of meeting the readers' needs?

Statement of Thesis

When you have finished answering the above questions, read them over and, based on what you have said, write a statement of thesis. (Refer to Chapter 2 if you need to refresh your memory.) The statement need be only two or three sentences. If you have trouble coming up with one, you need to give more thought to the problem-purpose-results connection.

General Outline

When your thesis has solidified, write it at the top of a separate sheet and then prepare a general outline under the three major categories labeled below. Follow the sequence indicated.

1. Under "evidence," list the main points you must bring out in support of your thesis.
2. What background information will the readers need in order to understand this evidence? List under "introduction."
3. Was the objective reached? What main conclusions can be drawn from the evidence? List under "evaluation."

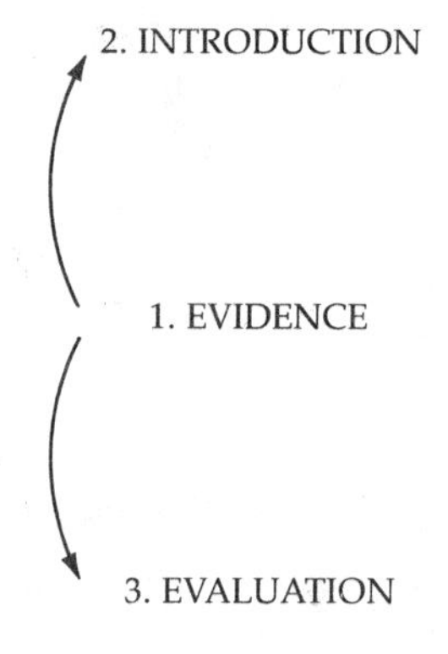

Detailed Outline

Next, prepare a detailed outline on a separate sheet as follows:

1. Condense the statement of thesis into a report title. Put title at top of outline. (You can also use the thesis statement later to help you write an abstract.)
2. Write the main heading INTRODUCTION. Under it, write brief descrip-

tive notations of: the problem, the objective, and the other points you brought out in the general outline.

3. Examine the points of evidence in your general outline. What method of developing these points will best meet the readers' needs? (Histori- cal? Categorical? Order of importance? Other?) What main topical headings does the method you have chosen suggest? Write these as main headings in your outline leaving space between. (They replace the general heading "Evidence.")
4. Review the notes you compiled during your investigation. Determine the details you should include by rating each detail as primary (this concerns graphic aids as well as prose), secondary, or irrelevant, ac- cording to the statement of the thesis. List the primary details under the appropriate main topical headings, *following whatever order best suits the subject matter.*
5. Write the main heading CONCLUSION. Under it enter the main conclu- sions you listed in your general outline. Put them in an order that par- allels the development of evidence.
6. What general conclusion should the reader draw from the detailed conclusions? Include it as a separate entry.
7. Do you have any recommendations or suggestions? If so, include un- der a new main heading.
8. Finally, what secondary material might be helpful to your reader? If you have any, enter it under the heading APPENDIX.

A sample outline of an actual report is given in Illustration 11–1.

Writing the Draft

The purpose of the rough draft is to take you from notes into prose *as quickly as possible.*

1. Use the entries in your outline as "talking points," just as you would if they were notes for a talk.
2. Try to "talk your way through" at least a major section at a time. Do not concentrate on grammar, word choice, or punctuation. Strive for continuity. Develop the story smoothly and logically from point to point.
3. Do not spend time rewriting sentences or otherwise polishing. That comes later.

Revising the Draft

The following points are suggested as a general procedure to follow. The check list that follows describes specific items.

1. Do not begin your revision the moment you finish your rough draft. Get away from it for a while.
2. Reread your draft all the way through before you actually begin to recast sentences or otherwise alter major pieces of prose. Do, however, make marginal notations of apparent discrepancies and errors as you read along.
3. If at any time you have to slow down in your reading or have to reread a passage, mark it for revision.
4. Reread each spot you marked in the first reading. Place the error in context; then determine whether you need to rephrase, delete, or add material. Correct accordingly.
5. Whenever errors are bunched, you may be able to revise more easily by rewriting the entire subsection or paragraph. Try to salvage the original first, but if you don't gain headway, write another version.

A CHECK LIST FOR EDITING YOUR WRITING

Check your writing in three areas:

1. Organization of the subject matter
2. Composition (grammar, punctuation, spelling, etc.)
3. Mechanics of format

Don't be alarmed by the length of the following list. Some points in the last two areas are fairly trivial and you would probably check them anyway.

Organization

1. Are all elements (title page, abstract, etc.) present? (Requirements differ according to report type.)
2. Are the elements in the proper order?
3. Is the purpose of the investigation clearly stated at the outset? The problem defined?
4. Is sufficient background material provided?
5. Does the presentation of evidence follow an order best suited to the reader's needs?
6. Do important points stand out? (Not buried under a mass of secondary details or data.)
7. Should material be eliminated or moved from the body to an appendix? (Details of procedure, derivations of equations, data sheets, etc.)

8. If specific questions are asked in the assignment, are the answers easily found? In the order in which the questions were asked?
9. In describing a concept or device, unfamiliar to the reader, does the passage begin with the function and purpose of the whole before presenting the parts in detail? (A circuit schematic or list of apparatus is not satisfactory here.)
10. From the evidence presented, are the conclusions valid?

Composition

1. Errors at the paragraph level:
   - Do all paragraphs have a topic sentence near the beginning?
   - Does each paragraph have only one topic?
   - Is there a clear and orderly development of the topic?
   - Do any paragraphs need more transitional words to tie together related ideas?
   - Are some paragraphs too long? (Most should be under half a page.)
2. Errors at the sentence level:
   - Are the subject and verb near each other?
   - Are modifiers near the words they modify?
   - Are subordinate thoughts written in subordinate grammatical constructions?
   - Are parallel thoughts in parallel constructions?
   - Does verb agree with subject in number? Pronoun with antecedent?
   - Are sentences punctuated correctly? (The troublesome marks are comma, semicolon, colon, hyphen, dash.)
   - Does every pronoun stand for something that has been expressed? ("It" and "this" frequently are vague.)
3. Errors at the word level:
   - Spelling correct? (This includes proper hyphenation.)
   - Meaning correct? (Often the wrong word is chosen; e.g., "imply" for "infer.")
   - Abbreviations standard?
   - Jargon, slang, clichés avoided?
   - Short, familiar word used instead of long, ornate?
   - Deadwood eliminated (roundabout expressions, "it . . . that" constructions, redundancies, unnecessary qualifiers)?
   - Do any technical words need to be defined for the reader?
   - Are names, titles, symbols consistent between text and figures or tables?

Mechanics

1. Pages numbered?
2. Figures and tables strategically located? Proper titles, labels? Referred to in the text?
3. Sufficient number of headings?
4. Do headings clearly show the relative weights of the sections they introduce? Typography and page positions consistent?
5. Footnotes executed properly?
6. Appendix material referred to in text?
7. Equations indented? Numbered? Symbols defined?

TESTS FOR EXPRESSION AND ORGANIZATION

You can test the adequacy of your main title and abstract by giving the abstract, without title, to a colleague unfamiliar with the project being reported. Ask your friend to read the abstract and write an original title. Compare this title with yours. The phrasing does not have to be exactly the same, but the two should agree in scope and emphasis. If they do not, then either your abstract is faulty or your title does not accurately represent your report.

You can test the validity of your conclusions by having someone with approximately the same background as your intended reader's read your report up to the section containing your conclusions. Ask this person not to look at your evaluation, but first jot down whatever conclusions are justified by the evidence you have presented and then to read your conclusions and comment on any disagreements.

Any technical description you may have had trouble writing also should be double-checked. Ask the same person who tested your conclusions to recount the passage in his or her own words. If your meaning has been misinterpreted, go over the written version with your friend and correct whatever was misleading. This test will take only about ten minutes, and it may well save your having to answer a lot of questions later. Be sure to try it whenever you must describe the operating principle of a new device; this type of technical description is a constant troublemaker.

If the reviewer of your manuscript comments that the organization is faulty (but cannot pinpoint what is out of place), make an outline of the manuscript. If you made one before you began to write, perhaps by altering it as you wrote you might have thrown the basic structure off balance. To check quickly, concentrate on the paragraphs. List, in note form, all the topic sentences; then see if the headings you use group the topics properly. Any flaw in logical order will show up immediately.

CITING THE LITERATURE

The following technique is from a style manual written for mechanical engineering students at M.I.T. It represents the feelings of at least some technical people about how references can be handled more effectively. Unless the editorial policy of your organization or the publication you may be writing for dictates otherwise, you might try the suggested method.

> You are undoubtedly familiar with the custom of using superscripts to refer to literature sources contained in footnotes or in a separate list of references at the end of the text. Of the two systems, prefer the listing at the end if you have more than just a few sources. Footnotes distract the reader and therefore should be kept to a minimum. (It would be foolish to make a separate list just for two references, however.)
>
> Another system, not so well known but actually more helpful to the reader, is to omit the superscript and substitute the last name(s) of the author(s) and the date of publication. The list of references also is different, the date of the publication coming directly after the name of the author. The following references illustrate the system.
>
> FOR IN-TEXT REFERENCES
>
> For similar treatments of the definition of stress and strain and the equations of equilibrium, see Reynolds and Smith (1983), Chapter 4; and Richards (1982), Chapter 2.
>
> FOR THE LIST OF REFERENCES

Ridler, Philip	1982	*Pocket Guide to BASIC*, Addison-Wesley, Reading, Mass.
Atkins, Robert	1984	*13-cm Moonbounce Experiments*, QST, June 1984, Am. Radio Relay League, Newington, Conn.

> Dates are made prominent so that readers will see immediately how recent the information is.

A SPECIAL PROBLEM: THE PROGRESS REPORT

The progress report is perhaps the commonest piece of technical writing. But it often presents a special problem: how does one report progress when there is no progress to report?

The question is not academic. At one time or another during an investigation many scientists and engineers feel that they have nothing to say (or want to say). They interpret "progress" to mean "a positive development, toward a defined goal, that can be verified by fact or accepted theory." Such a rigid interpretation is bound to create trouble for a writer, because most preliminary evidence is tentative and most preliminary results are fragmentary.

If you will bear in mind that "progress" here connotes "interim" or "periodic," you will realize that no special demands are made on you, as a writer, to report only positive results. Indeed, both positive and negative results are expected in progress reports, since the reports establish a running record of the investigation for present and future reference.

To help in overcoming any reluctance you may have with regard to reporting failures or committing yourself without an airtight defense, consider these additional points:

1. Preparatory work, at any stage of an investigation, represents progress.
2. Interim failures are by no means final failures, and the reader will not mistake one for the other.
3. An interim conclusion is a tentative conclusion; it can be retracted without penalty. Just be sure the reader knows the basis on which it is drawn.
4. If a result or conclusion cannot be given, a statement of the problem and a description of the attack will satisfy the reader.

In conclusion, the progress report is, among people with mutual interests and goals, an indispensable means of exchanging current information. To withhold information because it is disappointing, tentative, preliminary, or unimaginative is to defeat the purpose of the report.

PROJECT FOLDER

In a group project, each member must be kept informed of the plans and progress of colleagues. Project conferences, briefing sessions, and progress reports are the customary methods of exchanging this information. Not so common, but extremely helpful for a small group, is the project folder.

In effect, the project folder serves as a central file for all information bearing on the conduct of the investigation. Each member of the group makes an extra copy of any communication he or she initiates relative to the project, and files the copy in a predetermined division of the folder. Notes on consultations, brief minutes of meetings, photographs and sketches, and copies of relevant papers and reports from the literature are other items usually included.

One person can easily serve as coordinator and custodian. The folder may be kept in the library, in the group leader's office, or in any location convenient to the group. But no one should be allowed to take it from this location, since the contents should be available to any member of the group at any time.

At the end of the project, the folder may either be discarded or stored for future reference. Some companies have found it to be invaluable evidence for their patent department.

Note: If the members of the group use a word processor, the written segment of the project could be maintained in a storage file, with notations to that effect in the folder.

A PERSONAL JOURNAL

You probably have discovered that it's definitely helpful to keep a laboratory notebook, particularly when the time comes for writing a laboratory report. Keeping a personal journal is really an extension of that technique of information-gathering, but it has the distinct advantage of privacy (unless you decide otherwise). In Chapter 11, I mentioned other reasons why writing in a journal helps the creative process. But you don't have to take my word for it. Ask senior scientists and engineers in research for their opinions — I'm sure they will attest to the importance of the device. Think how grateful modern scientists are that Leonardo da Vinci kept a profusely illustrated journal!

HELPFUL REFERENCES

Probably the one device that will aid you the most as writer is a set of practical references. The collection does not have to be extensive, nor do you need to have every volume at your fingertips.

If your company or college department has a style manual, this guide and a good dictionary may be all that you will need as working references for most of your writing. Occasionally, however, you will want additional help, usually when you are working on an important, perhaps even critical, report.

To meet such emergencies, you should select from the literature of writing and editing those references which best fit the type of writing you have to do, and which you can use easily. Start with the titles in your company or college library; later, you can purchase any you find especially useful. Chapter 14 lists and evaluates titles from the fields of most interest to students, writers, editors, and teachers of technical information.

Chapter 13

Word Processing: A Versatile Tool

"I guess I could have pulled it off without a word processor, but I would have had a lot less sleep."

"I'd been told by friends that, as course deadlines neared, the computers were so overloaded that it took twenty minutes to get a command through. That didn't appeal to me."

Comments by two Harvard students on the use of a word processor to write a thesis.

PROLOGUE

Today, all technical people know something about computers, many own their own computers, and most are familiar with how computers function as word processors. But actual use of word processing as a personal writing tool is yet not widely established in the technical community.

Since personal computers have now invaded our schools and colleges and every student has an opportunity to use a word processor, the technical writing classroom is the logical place to develop this versatile new technique.

No one enjoys the endless struggle of revising a manuscript. Some people, however, have a built-in resistance to this chore that they can turn on at will and thus avoid the task entirely. The rest of us aren't so lucky.

Until recently, there seemed to be no practical way to convince students (and some teachers, for that matter) that a willingness to revise should be made a way of life. Fortunately, revision may be a much easier task from now on, thanks to the computer.

I am not one to be impressed by anything that poses as a quick fix or that claims it can make a writer out of a clod. But I am sold on word processing. Once the novelty and apprehension of operating a new system has worn off, one's attention can be fully centered on creating the message.

The purpose of this chapter is to encourage teachers of technical writing and their students to learn the technique of word processing without delay. The emphasis at the beginning is on background; anyone familiar with the basics of word-processing systems and operation, however, can skip to the next section, "Procedure." Software and hardware are mentioned only where a general understanding of computers is necessary.

BACKGROUND

A Working Definition of Word Processing (WP)

Word processing is the technique that uses the speed and adaptability of an electronic digital computer to accomplish the numerous *mechanical* operations involved in writing, revising, and printing a manuscript. In everyday terms, the operations the computer performs are:

1. Assembling the English words and data it receives via a keyboard into structures and formats specified by the user (for example: sentences, paragraphs, sections, chapters, line length, page size, tab, etc.).
2. Displaying the text, as it is being typed, on the screen of a video monitor.
3. Storing the text so that it can be recalled at will for revising.
4. Rearranging, centering, deleting, inserting, or transferring words, sentences, and paragraphs (or bits thereof), as specified by the user and then justifying all lines on a page.
5. Providing hard copy of the revised manuscript from a printer controlled by the computer.

These operations are basic to all computer systems that are used for WP. Other operations, of course, are possible; the options depend on the size of the computer, especially its memory, and the software it uses.

A Word About Word-Processing Systems

Word-processing systems come in a variety of shapes and sizes. Some large companies have multi-user systems in which many terminals tie into a dedicated main computer; others use word-processing software to operate large, general-purpose computers. Of wider use, especially in small businesses and in classrooms, are the so-called personal computers — microcomputers that are designed for desktop use and packaged as integral models or modular systems.

With the appearance of the personal computer, most U.S. colleges and universities have installed computer word processors in their writing classes and writing program centers. Machines for general student use also have been made available in limited numbers in common rooms and student centers. Harvard, for example, now has coin-operated computers in its freshman union and science center, as well as in many of its residential houses. Availability is still a problem for students, especially as the deadlines for courses near. M.I.T. is struggling with a similar problem and is now purchasing a large number of personal computers and reselling them to students and staff at a greatly reduced price. These problems, arising from the overnight demand for machines, should abate in the near future.

The Basic Components

A word-processing system consists of the following basic components:

- *Keyboard*. Similar to that of a regular typewriter but with extra keys that initiate special functions.
- *Video monitor*. Similar to a TV screen, for viewing the text as it is being typed in at the keyboard. (A regular TV also can be used with certain models of home computers.)
- *Processor*. The central processor is the heart of the computer. It is used to coordinate, execute, time, and control all of the operations that are necessary for the completion of each writing, editing, and printing task.
- *Software*. A program that serves as a set of instructions for the processor to follow when it is asked to work on each task. A word processor is only as good as its software.
- *External storage*. A device, such as a floppy or hard disk, for storing the processed text so that it can be recalled at will for referencing and editing.
- *Printer*. A peripheral device that provides hard copy of the text when the revisions have been completed and, if the memory capability exists, during the processing when desired.

With the exception of the software, all of the above are standard components. The software is special. Software programs are sold as separate packages by the computer manufacturers and by independent companies that specialize in software production. Software for word processing is available for all popular makes of personal computers.

A Word-Processing System for Student Use

As mentioned earlier, college writing programs are now adding word processors to their resource centers for use by students in writing courses and by those involved in major writing projects, such as theses. Ideally, such a facility would enable students to:

- Tie into the computer from a dormitory terminal to get course assignments, reference material, instructions, and computer mail
- Use the computer at the center for processing writing assignments, from writing the rough draft to printing hard copy
- Use a remote terminal at the center to tie into the main computer(s) for library files of science and engineering research and other data pertinent to an ongoing project
- Through the tie-in with the mainframe, be able to get rapid printing of long reports and theses, especially when multiple copies are needed*

Again, ideally, the WP facility would also provide a graphics printer that would be available on a time-sharing basis and a modem for tie-in with computers at off-campus R&D laboratories run by the college, so that students could keep up to date on the progress of an investigation. Although the hardware and software are available today for all these services, it will be a while before a facility such as the one described is commonplace throughout all our higher educational institutions. But much can be done even when equipment and services are limited, as the remainder of this chapter will show.

PROCEDURE

Learning the Ropes

1. You don't need to know how a computer operates in order to use a word processor, but it's not a bad idea to glance through the literature that comes with the hardware. Do your reading where you can examine the equipment as you read. Knowing a bit about how your particular system functions can make your experience more satisfying.
2. The manual that comes with the software gives all the commands. It probably will take about a week to become familiar with these commands and another week before their use becomes second nature to you.

* At M.I.T., students may purchase a program that allows a P.C. to access one of M.I.T.'s mainframes and obtain copies from a laser printer at low cost.

3. Don't wait until you have learned all the commands to begin practice. (Most software begins with a lesson on procedure, telling you on the monitor what steps to follow.) Take several sessions just to get used to the equipment. There's no harm in having fun, too. (I learned to play some music with my computer.) Remember, you can't hurt the equipment by typing into it, so don't worry about making mistakes. In fact, it's O.K. to goof, because then you can find out how easy it is to correct errors.

Suggestions for Getting Started

1. Begin actual processing with a *short* piece, such as a memo or letter. Work from a brief outline this first time. You can create it on the spot at the top of your monitor screen. Directly under the outline, start your letter. Type out the whole piece as a first draft. When the screen gets full and the text begins to scroll, you will lose only the outline as you use it.
2. Revise your letter into a polished copy. Read the text through from beginning to end as a test of content and continuity. Have you included all the necessary information? Does the message flow easily? Add any information, using either your insert command or block-transfer technique. Reread. If still not satisfactory, try again. That's what a WP is for.
3. Now go back to the beginning and examine each sentence, polishing where necessary. The computer will take care of the mechanics.
4. Read over your final version. If you like it, you're ready for a printout. Use whatever commands your system requires to initiate the printing of hard copy.
5. Although you do no harm to the computer when you make mistakes in typing a command, it's a good idea to be precise when you ask the computer to do something. (After all, you expect *it* to be precise.) So try to reproduce a command exactly as it is printed in the manual; even the wrong spacing could call up an error statement on the monitor.

Suggestions for Writing the Draft of a Report

In processing your memo or letter you went through four steps in one sitting: outlining, writing the draft, editing, and printing. This procedure works fine for a short piece, but it isn't practical for something longer, such as a technical report or a journal article. You can, however, produce a quick, satisfactory draft if you use a standard textual organization or, if not, have the general coverage of your report clearly in mind. These are the steps for this task:

1. List vertically from top to bottom, on the left side of the monitor, descriptive headings that represent the content of the major sections of your report. Leave four or five lines of blank space between each heading. You may need more than a "page" to complete this step.
2. Under each heading, summarize in a short paragraph what information you think each section should contain. Don't be fussy about the wording, and if you make a typo, don't stop to correct. Try to complete this step in no more than ten minutes.
3. With the above completed, you now have a mini draft of your report. Take a few minutes to go over what you have typed on the screen. Concentrate on the message. Use the WP to add, delete, or transfer information.
4. You are now ready to store the mini draft. Prepare the text in whatever system your computer requires. (I suggest giving the piece an identifying name; e.g., DRAFT 1 BIOLOGY REPORT, and numbering the headings or even the lines so that you can change parts if you wish.) Initiate the commands necessary to store the above.
5. When storage is completed, command a verification of what is stored, just to make sure that everything went O.K.

The next stage of the process is to develop the text under each heading of DRAFT 1. I suggest the following procedure:

1. Type in the commands to read and display the first section of the draft you just stored. With this at the top of the screen for reference, flesh out the short paragraph with details. One way to do this is to make each sentence the topic sentence of a new paragraph and then to develop that topic. Do this as rapidly as you can, and again, concentrate on content, not expression. As you enlarge the information, add any subheadings that will help tie ideas together or provide transition.

   Follow this procedure through all the sections. If you get stuck for an idea, don't stare at the screen; run the cursor down to the bottom and try out a sentence there. This technique relieves some of the pressure and lets you express yourself until you feel that what you've said tells the right story. You can check the spot again at revision time.
2. You have now completed a draft of the text of your report. Store it for use at the next stage. I recommend that you also get a printout.

   Before you begin editing the draft, take a breather so that you'll be able to have a fresh look at what you've written. If you wish, you can give some thought to ancillary matters, such as planning how to incorporate any graphics you might wish to use and preparing the service material (front matter and appendices) that will be needed.

Advantages of Using a Word Processor for Creating a Draft

Apart from the mechanical advantages of deleting unwanted material, inserting new material, and transferring material from one spot to another, a word processor does not require a retyping when a revision is made and, with a printer, it enables the user to get fast hard copy of the text or even of parts of the text. There is definitely an advantage in being able to see what the finished product will look like. And a hard copy of a draft can be shown to others for comment. (Unfortunately, not many students have ready access to a printer. They therefore lose the advantage of being able to see hard copy of a draft before it is time for revision. Teachers of technical writing should do all they can to arrange a time-sharing scheme of a printer for their classes whenever they give a major writing assignment.)

Other advantages include: avoiding all the clutter of scraps of crumpled paper and filled wastebaskets; eliminating the chance to lose inserts or revisions; not having to worry about leaving enough space in the right place for a graphic aid; and perhaps the best advantage, the joy of having a clear video copy to work on — no ugly deletions or inserts which, in themselves, often tend to block creativity. In fact, one student reported to me that he was able to produce three drafts of a thesis, using word processing, in the same amount of time it took him to do a single draft of a course paper of comparable length, using the conventional longhand/typewriter method.

Revising the Draft for Coverage and Organization

In reviewing the draft for revisions, concentrate first on the overall impact of the message. Read through the entire text at your normal reading speed. Do you need to rearrange the presentation at any point? Transfer any block of information? Delete or add any data? Reword or insert a heading? If there's any doubt in your mind, experiment. Move paragraph three after paragraph five, for instance. If you don't like it there, move it back. The computer will do all the typographical work for you. Correct any obvious errors in spelling, punctuation, and grammar, but don't go looking for errors at this point. There's no sense spending time polishing the wording of a paragraph before you have decided definitely that what it says is what you want.

If you run into real trouble, try this device. Remember the mini draft (Draft 1) you made? You can use it again to help you check the content. If you made a hard copy, get it out; if not, make one now. The procedure is to read over the section of your final draft that's causing a problem, concentrating on the message that comes through. Then stop for a moment and ask yourself to define the thesis of this section, as worded. Compare your answer with the statement you made in the mini draft. If the two jibe, you are on the right track, but something is wrong with the way you express yourself. If the two do not come close to agreeing, how do they differ? Why? Settle this dilemma before going on to the next section. Easier said than done, you say. True, but you still have a better chance of licking the problem with a word processor to help you with the mechanics than you would if you were using a passive (and most uninteresting) blank piece of paper!

(Before reading the next section on revising the draft, you might wish to skim the check list for editing, covered in Chapter 12.)

Revising the Draft for Grammar and Style

Let's assume that you're happy with the content of your report and the way the material is organized. You can now concentrate on polishing your presentation. As you read through the text on the monitor, underscore (or otherwise mark) any word or passage you hesitate on or think you should change. When you finish the reading, go back to the beginning and work on each place you marked. I recommend this method rather than revising as you read because you need to see what follows a possible trouble spot so that correcting one trouble will not cause another. If you can't think of a revision immediately, go on to the next marked item.

When you return to any trouble spot, if you still aren't happy with a revision, try this approach:

1. Start with *only* the bare subject, verb, and object of the sentence. (Cut *all* qualifiers.)
2. Do these key words represent your thoughts accurately? If not, your trouble is basic. Work on these elements *alone* before you begin to add qualifiers.
3. When you are satisfied with the basic sentence, add the qualifiers *one by one*. Read the sentence after each addition to make sure the sense of the message has not changed. This technique usually works because you see a complete sentence at each stage of the reconstruction and can appraise the meaning in a methodical manner (another advantage of using WP).

The computer will not think of words for you. But it will let you try all sorts of sentence combinations: beginning a sentence with the subject, beginning with an introductory phrase, using the active voice, using the passive voice, making two short sentences out of one long one (or vice versa), inserting a heading. You can try making a list instead of spelling items out in a sentence.

Finally, word processing can help you by automatically numbering the pages, justifying the right margin like that on a printed page, and checking your spelling (if you but provide it with a "dictionary"). More sophisticated programs can automatically create a table of contents, provide you with an index if you just tell it what to look for, and produce a reasonably good abstract if you supply it with sufficient criteria with which to make selections. Also, there are now so-called integrated software packages that allow you to add graphics or tables of numerical data from other programs directly into your written text. There are even some programs on the market that, according to their creators, will let your computer serve as an editor, eliminating awkward phrases and needless repetition from your writing. I have not seen these programs in operation, so I cannot vouch for their capability. Before buying one of them, I would want to know if it supplies an "unawkward" phrase for the awkward one it eliminates. Also, how does the program decide when repetition is needless? So be wary of any product that claims to make subjective judgments on questions of style.

SUGGESTIONS FOR A FUTURE PROJECT

I would like to encourage those of you who have some experience in writing computer programs, or plan to take a course in programming soon, to try your hand at writing one for your own use in word processing. Here are a few possibilities:

- A program to check the spelling of the special technical terms used in your profession or field of study
- A program that will automatically format a special type of report that you use frequently; for example, a laboratory report
- A program that will print out, in the order of their appearance, all sentences that contain more than thirty words. (These are considered long sentences for technical writing and should be checked out for clarity.)

These programs should not be overly complicated; I know of several homegrown ones. The one below, however, is difficult if you really want to be challenged:

- A program that will examine passages to determine if they meet the criteria for proper pace given in Condition 1 of the Guide for Control of Pace in Chapter 7.

In conclusion, you may have figured out from what I've said that this chapter was written and revised with the aid of a word processor. Although the time it took me to process the text cannot be compared meaningfully with that for the other chapters, where I used the conventional longhand/typewriter method, I definitely felt that I was working more smoothly and efficiently, and I do not hesitate to recommend WP to writers and students in all fields. Evidently, others feel the same way: Harvard Business School now requires all entering students to use personal computers as part of their regular class preparation.

EPILOGUE

The states of the art of hardware and software are advancing so rapidly that advice given today on procedure for any computer operation may not be worth much tomorrow. However, the procedure for word processing considered here applies, not to mass production of office paperwork, but to an individual's writing effort. It therefore should still be valid even with faster and more sophisticated systems. Greater speed and larger memories are not essential, however, to satisfactory word processing. Where they will help is with the complex tasks, possible now only with the largest machines: translating, abstracting, and indexing. The writer will still be involved in most of the tasks we have discussed here.

Chapter 14

Helpful References on Technical Communication

PRIMARY REFERENCES

Technical Writing

Andrews, Deborah C., and Margaret Blickle. *Technical Writing, Principles and Forms*. 2d ed. New York: Macmillan, 1982. A classroom textbook for beginning writers, with instructor's manual and built-in exercises. Covers audience analysis, collecting and screening information, verbal and visual presentation. Hardback.

Jay R. Gould, ed. *Directions in Technical Writing and Communication*. Baywood Technical Communication Series. Farmingdale, N.Y.: Baywood Publishing Company, 1978. The first in a series of paperbacks. Contains an excellent collection of articles by authorities on all aspects of the field, including oral presentation and graphics. Of particular interest to experienced writers who wish to broaden their perspectives.

Brusaw, Charles T., Gerald J. Alred, and Walter E. Olin. *Handbook of Technical Writing*. 2d ed. New York: St. Martin's Press, 1982. A reference book that combines points of grammar, punctuation, vocabulary, and usage, and types of written communication in an easy-access format. Paperback.

Houp, Kenneth W., and Thomas E. Pearsall. *Reporting Technical Information*. 2d ed. Beverly Hills, Calif.: Glencoe Press, 1975. A textbook of the basics but with some extras: a comprehensive listing of science and engineering reference books, a selected bibliography, a built-in handbook on grammar, and a main section that illustrates how to apply the principles presented earlier. Paperback.

Levine, Norman. *Technical Writing*. New York: Harper & Row, 1978. A handy paperback edition divided into three parts: technical explanation, technical mechanics, and technical reporting. The last part contains a good presentation of special formats. Could serve as a self-instruction text.

Smith, Richard W. *Technical Writing: A Guide to Manuals, Reports, Proposals, Articles, etc. in Government and Industry*. 5th printing. New York: Barnes and Noble College Outline Series, 1968. Like all the other publications in the COS series, this one gives buyers more than their money's worth. It is not a hodgepodge of grammar, usage, and technical writing, but sticks to info on the field. Even has chapters on films and newspaper writing. Paperback.

Scientific and Engineering Writing

Day, Robert A. *How to Write and Publish a Scientific Paper*. Philadelphia, Penn.: ISI Press, 1979. A basic how-to guide, to be used with style manuals published by scientific societies. Has good practical suggestions for would-be authors. Paperback.

Kirkman, John. *Good Style for Scientific and Engineering Writing*. London: Pitman Publishing, 1980. Written specifically for engineers and scientists in this country as well as the U.K. Covers choices of vocabulary, phrasing, and sentence structure. Examples taken from actual writings by engineers and scientists. Stresses the achievement of overall readability combined with accuracy. Includes a chapter on writing for users of English as a foreign language. A fine book. Paperback.

Institute of Electrical and Electronics Engineers. *A Guide for Better Technical Papers*. Edited by Craig Harkins and Daniel L. Plung. New York: IEEE Press, 1982. A volume in the selected reprint series by IEEE. Well worth having. Contains articles on "Getting Started," "The Rhetoric of Papers and Articles," "Tricks of the Trade," "Some Research Results," and "Following Through." The one on "Tricks of the Trade" is worth the price alone. Comes in paperback or clothback, distributed by John Wiley & Sons, New York. Order by number: paper PP01537; cloth: PC01529.

Society for Technical Communication. *Proceedings of the 29th International Technical Communication Conference*. Washington, D.C., 1982. A worthwhile collection of papers on many subjects, including an excellent one on "The Engineer as Technical Communicator" by John Simons.

Technical Editing

Bennett, John B. *Editing for Engineers*. A Wiley-Interscience Book. New York: John Wiley & Sons, 1970. One of the Wiley series on human communication. An excellent guide on how to edit, what to edit, and how to work with writers. Has stood the test of time. A must for the engineer-editor or someone wishing to become a technical editor. Hardcover.

Editorial staff of the University of Chicago Press. *A Manual of Style*. 13th ed. rev. and exp. Chicago: University of Chicago Press, 1982. Updated many times since it was first published in 1906, this is a primary reference book for authors, editors, publishers, and printers. Covers planning a book, rules for preparing copy, hints to authors, glossaries of technical terms and symbols, and specimens of type. Hardcover.

Usage and Style

Nicholson, Margaret. *American English Usage*. New York: Oxford University Press, 1957 (hardback); New American Library, 1967 (Signet paperback Y3582). An edited version of H. W. Fowler's *Modern English Usage*, which also should be on every reference shelf. (Nicholson really changes very little.) Written with gusto and witty erudition. Particularly recommended are: "Formal Words," "Elegant Variation," "The Split Infinitive," "Love of the Long Word," and "Avoidance of the Obvious." Suggest purchasing the paperback of Nicholson version and putting the Fowler hardcover on your Christmas list.

Strunk, William, Jr., and E. B. White. *The Elements of Style*. New York: Macmillan, 1959; 3d ed. 1979. Originally a composition handbook used at Cornell; revised and added to by E. B. White. Became a best-seller because of White's popularity and reputation as a guardian of usage. Although old hat now, it still should be on every reference shelf. It is short, to the point, witty, and inexpensive, and can be read in one evening. Available in paperback and hardcover.

Williams, John M. *Style — Ten Lessons in Clarity and Grace*. Glenview, Ill.: University of Chicago Press/Scott, Foresman & Co., 1981. This book is for writers who no longer worry over their spelling and verb forms but are concerned with writing prose that is clear, direct, and even graceful. It's more than a rhetoric, more than a handbook. One chapter, for example, "A Touch of Class," offers a distinctly different approach with its emphasis on "grace notes."

Graphics

MacGregor, A. J. *Graphics Simplified*. Toronto: University of Toronto Press, 1979. This is the most recent book I could find on graphics. It is a good review of the field for beginners. If supplemented with articles on special applications of graphics from the Society for Technical Communication's publications, the package would be comprehensive and extremely useful.

Oral Presentations

Institute of Electrical and Electronics Engineers. *A Guide to Better Technical Presentations.* Edited by Robert M. Woelfle. New York: IEEE Press, 1975. A volume in the IEEE Press Selected Reprint Series. Includes articles on planning and preparation, visual aids, delivery techniques, multimedia presentations, and motion pictures. Clothbound and paperbound distributed by John Wiley & Sons, New York.

Proposals

Whalen, Tim. *Preparing Contract-Winning Proposals and Feasibility Studies.* New York: Pilot Books, 1982. Covers the entire picture of the process, beginning with the initial decision; and on to proposal strategies, understanding the client's point of view, evaluation check lists, writing résumés, production techniques for text and artwork. Paperback.

Dictionaries

Sippl, Charles J. *Computer Dictionary.* Indianapolis, Ind.: Howard Sams, 1982. Intended to acquaint nontechnical people with computer jargon. Paperback.

The American Heritage Dictionary of the English Language. Boston: Houghton Mifflin Co., any edition from 1969 on. A fine general source with many new technical words.

SUPPLEMENTARY REFERENCES

The Writing Process

Baker, Sheridan. *The Practical Stylist.* 5th ed. New York: Harper & Row, 1981. Contains a composition text and a writer's handbook. Emphasis on structure, tactics, how to conduct a research project, how to present evidence. Paperback.

Elbow, Peter. *Writing with Power.* New York: Oxford University Press, 1981. A new look at some old problems. Features several unique approaches to mastering the writing process: "freewriting," "loop-writing procedures," "the open-ended process." If you are not familiar with the Elbow approach, you really should inspect this book. Paperback.

Grammar and Composition

Leggett, Glen, C. David Mead, and William Charvat. *Handbook for Writers.* 8th ed. Englewood Cliffs, N.J.: Prentice-Hall, 1984. Sections on grammar, basic sentence faults, manuscript mechanics, punctuation, and effective paragraphs. Also covered: writing summaries, business letters, and research papers. In all, a comprehensive, easy-to-use handbook. Hardback.

Turner, Rufus P. *Grammar Review for Technical Writers*. rev. ed. New York: Holt, Rinehart and Winston, 1971. Covers the major trouble spots; uses examples from technical writing. Easy-to-use format. Paperback.

Word Processing

New journal articles, magazines, and books on the subject appear almost daily, so that any cited here might well be out of date before this edition is published. I will take a chance with one book, however, because it says what all the others do in a much more readable manner.

Zinsser, William. *Writing with a Word Processor*. New York: Harper & Row, 1983. A humorous and informative book about the problems, some imagined but most very real, that a nontechnical (in fact, wary-of-anything-mechanical) writer encounters when he or she enters the world of computers and finally comes out singing the praises of word processing.

Electronics

Concepts of Electronics. Book I. Heathkit Educational Series. Benton Harbor, Mich.: Heathkit-Zenith, 1981. This is a self-help book designed to give a thorough foundation in the basic principles of electronics for the nontechnical person and progresses to intermediate principles for those who wish to learn more. I suggest this book for any technical writer or editor who has no background in electronics but who will be working in the electronics field. Paperbound.

Fun-to-read Pieces

Eschholz, Paul, Alfred Rosa, and Virginia Clark, eds. *Language Awareness*. 3d ed. New York: St. Martin's Press, 1982. This is an excellent collection of articles by well-known authors. Included are: "Language on the Skids" by Edwin Newman and "Gobbledygook" by Stuart Chase. Other titles: "The Language of Advertising," "The Sounds of Silence," and "A Brief Lexicon of Jargon for Those Who Want to Speak and Write Vaguely and Verbosely." The collection makes good reading and will provide an abundance of passages to use in the classroom. Paperback.

Baker, Robert A., ed. *A Stress Analysis of a Strapless Evening Gown (and Other Essays for a Scientific Age)*. Garden City, N.Y.: Doubleday, 1969. (A Doubleday Anchor Book, #A648G) This collection is a delightful reassurance that there are scientists eager to keep a sense of proportion, dedicated to the proposition that no one can survive the rigors of our age without a sense of humor. Included are: "The Chisholm Effect," "Cosmic Call," "On the Nature of Mathematical Proofs," and "Parkinson's Law in Medical Research," as well as the famous "Stress Analysis . . ."

Index